The Mutual Assistance and Solidarity Clauses

European University Studies

Europäische Hochschulschriften

Publications Universitaires Européennes

Reihe II **Law**

Series II Rechtswissenschaft

Série II Droit

Volume/Band **5604**

Antonio-Maria Martino

The Mutual Assistance and Solidarity Clauses

Legal and Political Challenges of an Integrated EU Security System

PL ACADEMIC RESEARCH

Bibliographic Information published by the Deutsche Nationalbibliothek
The Deutsche Nationalbibliothek lists this publication in the Deutsche Nationalbibliografie; detailed bibliographic data is available in the internet at http://dnb.d-nb.de.

ISSN 0531-7312
ISBN 978-3-631-65258-9 (Print)
E-ISBN 978-3-653-04341-9 (E-Book)
DOI 10.3726/978-3-653-04341-9

PL Academic Research is an Imprint of Peter Lang GmbH.
Peter Lang – Frankfurt am Main · Bern · Bruxelles · New York · Oxford · Warszawa · Wien

This book was peer reviewed prior to publication.

www.peterlang.com

For my mother

Preface

The principle of solidarity is a fundamental principle of the European Union and references to 'solidarity' can be found in various provisions in the EU Treaties. The Treaty of Lisbon introduces a new provision, the 'Solidarity Clause', which obliges the Union and its Member States to act in a spirit of solidarity in the event of a terrorist attack, a natural or a man-made disaster. It should also enable the Union to prevent and respond to terrorist threats by mobilising all the necessary military and civil instruments.

Long before formal discussions within EU Council structures, the legal and political challenges of this new provision were discussed within an informal group composed of high-level officials of the Ministries of the Interior and Justice of several Member States. The 'Strategic Circle Internal Security', founded in 2010 on an initiative of the Austrian delegation in the Art. 71 TFEU Committee (COSI), paved the way for formal discussions within Council structures. These formal discussions started in 2011 and aimed at preparing the drafting of the proposal for a Council decision on the implementation by the Union of the Solidarity Clause. In 2014 this proposal, which was jointly submitted by the Commission and the High Representative, will be adopted by the Council.

I have been lucky enough to be involved in the discussions on the interpretation of Article 222 TFEU from the very beginning, which inspired me to write my Master's thesis on the Solidarity Clause and the thematically related Mutual Assistance Clause enshrined in Article 42 TEU. For this purpose it came in more than useful that I could choose as supervisor for the thesis one of the major experts with regard to both Clauses, Professor Gerhard Hafner, to whom I am indebted for continued advice and assistance. I would also like to thank Professor August Reinisch and his staff at the Department for European, International and Comparative Law of the Faculty of Law at the University of Vienna for the excellent organisation of the Master's programme 'LL.M in International Legal Studies'. I am also immensely grateful to Wilhelm Sandrisser, Director for International Affairs in the Federal Ministry of the

Interior (FMI), for giving me the opportunity to follow the developments on the Solidarity Clause from the 'front row' and for stimulating my research on the topic. I also thank Ariane Holezek, Head of Department in the FMI, for inspiring me to apply to the Master's programme 'LL.M in International Legal Studies' and Franz Einzinger, Director General in the Federal Ministry of the Interior, as well as Elisabeth Wenger-Donig, Vice Director of the Federal Office for Immigration and Asylum, for their benevolent support of my participation in the above-mentioned Master's programme. I am also very obliged to Sue Norris for her meticulous revision of the text. Finally, my most profound thanks go to my parents for their continued and generous support and to my wife Petra for her amazing tolerance with regard to my often (leisure) time-consuming research activities.

I have endeavoured to state the law and the state of play of the negotiations on the joint proposal for a Council Decision on the arrangements for the implementation by the Union of the Solidarity Clause as of January 1, 2014.

Antonio-Maria Martino
January 2014

The Mutual Assistance and Solidarity Clauses

Article 42(7) TEU

If a Member State is the victim of armed aggression on its territory, the other Member States shall have towards it an obligation of aid and assistance by all the means in their power, in accordance with Article 51 of the United Nations Charter. This shall not prejudice the specific character of the security and defence policy of certain Member States.

Commitments and cooperation in this area shall be consistent with commitments under the North Atlantic Treaty Organisation, which, for those States which are members of it, remains the foundation of their collective defence and the forum for its implementation.

Article 222 TFEU

1. *The Union and its Member States shall act jointly in a spirit of solidarity if a Member State is the object of a terrorist attack or the victim of a natural or man-made disaster. The Union shall mobilise all the instruments at its disposal, including the military resources made available by the Member States, to:*
   (a) — prevent the terrorist threat in the territory of the Member States;
   — protect democratic institutions and the civilian population from any terrorist attack;
   — assist a Member State in its territory, at the request of its political authorities, in the event of a terrorist attack;
   (b) assist a Member State in its territory, at the request of its political authorities, in the event of a natural or man-made disaster.
2. *Should a Member State be the object of a terrorist attack or the victim of a natural or man-made disaster, the other Member States shall assist it at the request of its political authorities. To that end, the Member States shall coordinate between themselves in the Council.*

3. *The arrangements for the implementation by the Union of the solidarity clause shall be defined by a decision adopted by the Council acting on a joint proposal by the Commission and the High Representative of the Union for Foreign Affairs and Security Policy. The Council shall act in accordance with Article 31(1) of the Treaty on European Union where this decision has defence implications. The European Parliament shall be informed. For the purposes of this paragraph and without prejudice to Article 240, the Council shall be assisted by the Political and Security Committee with the support of the structures developed in the context of the common security and defence policy and by the Committee referred to in Article 71; the two committees shall, if necessary, submit joint opinions.*
4. *The European Council shall regularly assess the threats facing the Union in order to enable the Union and its Member States to take effective action.*

Table of Contents

Summary

The 'Mutual Assistance Clause' (Article 42(7) TEU)[1] and the 'Solidarity Clause' (Article 222 TFEU) were formally introduced by the Lisbon Treaty[2]. Whereas the Mutual Assistance Clause has models such as the NATO Treaty[3] and the modified Brussels Treaty of 1954[4], which established the Western European Union (WEU), the Solidarity Clause is a 'genuine' innovation of the European Union. Although the overall aim – assistance to a Member State affected by an extraordinary situation – is the same and the principle of solidarity underpins both provisions, there are legal differences and also possible overlaps between the two Clauses.

According to Article 5 NATO Treaty, an armed attack against one of the signatories is to be considered an attack against them all. Therefore, in line with Article 51 of the UN Charter, such an attack would give rise to the right to self-defence for all NATO Member States. Furthermore, Article V WEU states that the signatories are to support the attacked State with all the military and other aid and assistance in their power. The Mutual Assistance Clause in Article 42(7) TEU applies in cases of armed aggression by a third State and is notable for its brevity, simplicity and vagueness. With regard to its legal setting, the Mutual Assistance Clause is not the object of a separate article itself but is covered by Article 42 TEU. The Clause does not include any reference to implementing procedures or to operational mechanisms, nor does it give any role to EU institutions.

---

1 Official Journal of the European Union C 326/1, 'Consolidated Versions of the Treaty on European Union and the Treaty on the Functioning of the European Union', 26.10.2012.

2 'Treaty of Lisbon amending the Treaty on European Union and the Treaty establishing the European Community', UNTS I-47938.

3 'North Atlantic Treaty' of 4 April 1949, 34 UNTS 243–246, available at http://www.nato.int/docu/basictxt/treaty.htm.

4 'Treaty between Belgium, France, Luxembourg, Netherlands and the United Kingdom of Great Britain and Northern Ireland for collaboration in economic, social and cultural matters and for collective self-defence', UNTS I-304.

The Solidarity Clause is a unique horizontal provision striving to encompass and connect the intergovernmental 'Common Foreign and Security Policy' (CFSP) with supranational policy areas such as the 'Area of Freedom, Security and Justice' (AFSJ) and 'Civil Protection'. It also has a possible military dimension, which makes it even more difficult to clearly set the limits of its scope. The Solidarity Clause is, therefore, a provision which aims at interlinking internal and external security. Despite the fact that the competent Working Group of the European Convention (Working Group VIII) proposed the Solidarity Clause as a complement to the Mutual Assistance Clause, the drafting of the former raises a number of legal and political questions. Unlike the Mutual Assistance Clause, Article 222(3) TFEU foresees the necessity to adopt an implementing provision in order to establish the arrangements for the implementation of the Solidarity Clause by the Union. The draft Council Decision ('Joint Proposal') that should set out this implementation is currently being discussed within the preparatory bodies of the Council. Due to its rather vague wording and its (geographic) scope, however, the Joint Proposal raises further legal questions.

The Mutual Assistance Clause and the Solidarity Clause are binding on all Member States, which are obliged to act in a spirit of solidarity. Both Clauses do not limit the assistance nor the means that can be applied to assist the affected Member State(s). Despite these similarities, they are separate in legal terms as the one clause has a clear intergovernmental background whereas the other is of a supranational nature. Certainly the supranationality of the Solidarity Clause does not prevent it from potentially extending its scope to intergovernmental areas such as defence and foreign policy. The intergovernmentalism of the Mutual Assistance Clause is absolute as no involvement of Union institutions is foreseen, whereas in the Solidarity Clause the Union is not only a direct addressee of the solidarity obligation, but has also a role (albeit a marginal one) as regards the Member States' efforts in providing support to the affected Member State(s). Non-compliance with the obligation to assist arising out of Article 42(7) TEU does not result in any legal consequences. What is more, mandatory assistance is only binding on those Member States which are also NATO members, whereas the neutral and non-aligned Member States are, by and large, exempted from any obligation to assist. The Member State(s) that decide not to abide by the obligation established in the Solidarity Clause could incur in infringement proceedings before the Court of Justice of the European Union (CJEU). With respect to procedural issues, both Clauses are also very distinct: the Mutual Assistance Clause does not prescribe any procedure to be followed in the case of armed aggression; the Solidarity Clause refers to implementation arrangements which are still awaiting their final adoption. Even with regard to the assistance by the Member States, the Solidarity Clause refrains from limiting the freedom of the Member States and provides for a forum (the Council) where they are to coordinate themselves.

The following analysis focuses on the origins, scope, interpretation and technical aspects of the Mutual Assistance and Solidarity Clauses. It highlights the possible overlaps of both clauses and shows the wide range of competences theoretically conferred upon the Union and the Member States by the Solidarity Clause in particular. The paper gives also an outline on recent developments with regard to the so-called Joint Proposal which will implement Article 222(3) TFEU and set out the framework for Union action in a situation triggering the Solidarity Clause ('solidarity case').

I. Introduction

The Mutual Assistance and Solidarity Clauses were formally introduced by the Lisbon Treaty in December 2009 and are now enshrined in Title V, Chapter II, Section 2 'Provisions on the common security and defence policy', Article 42(7) of the Treaty on the European Union (TEU), and in Part V 'External Action of the Union', Title VII, Article 222 of the Treaty on the Functioning of the European Union (TFEU). Until now, they have only received limited attention from scholars and analysts, as well as from national governments and EU institutions. This may change in the near future as the European Council discussed in December 2013 the current status and possible future of the 'Common Security and Defence Policy' (CSDP) and the General Affairs Council should adopt a Council Decision on the implementation of the Solidarity Clause ('Joint Proposal') in June 2014.

These two clauses raise a number of questions regarding their scope and legal interpretation, their potential for further integration and their practical and technical implementation. Mutual assistance and solidarity are somehow closely related terms in European Union law. If we take a closer look at the current legal framework of the European Union we find several references to 'solidarity'. Apart from the Preamble to the Treaty on the European Union ("DESIRING to deepen the solidarity between their peoples while respecting their history, their culture and their traditions"[5]) there are several other references to 'solidarity' in primary law.

The Union "shall promote [...] solidarity among Member States" (Art. 3(3) TEU) and "shall contribute to [...] solidarity and mutual respect among peoples" (Art. 3(5) TEU). "The Union's action on the international scene shall be guided by the principles which have inspired its own creation, development and enlargement", inter alia the "principles of equality and solidarity" (Art. 21(1) TEU). In Part Three, Title V, Article 78(3) TFEU, relating to the

5 Official Journal of the European Union C 115/13, 'Consolidated Version of the Treaty on the European Union', 9.5.2008.

Area of Freedom, Security and Justice (AFSJ) we find, for example, a provision on the possible adoption of measures aimed at supporting a Member State in the event of a massive migratory influx:

> "In the event of one or more Member States being confronted by an emergency situation characterised by a sudden inflow of nationals of third countries, the Council, on a proposal from the Commission, may adopt provisional measures for the benefit of the Member State(s) concerned. It shall act after consulting the European Parliament."

However, Article 80 TFEU evokes the principle of solidarity and fair sharing of responsibility with regard to the whole chapter on border checks, asylum and immigration policies (hence, also applying to Article 78):

> "The policies of the Union set out in this Chapter and their implementation shall be governed by the principle of solidarity and fair sharing of responsibility, including its financial implications, between the Member States. Whenever necessary, the Union acts adopted pursuant to this Chapter shall contain appropriate measures to give effect to this principle."

Article 80 TFEU "is not a legal base in itself, but provides justification for the adoption of measures in future such as the European Refugee Fund"[6].

We also find the term 'solidarity' in other policy areas covered by the TFEU. According to Article 122 TFEU, "the Council, on a proposal from the Commission, may decide, *in a spirit of solidarity* [emphasis added] between Member States, upon the measures appropriate to the economic situation, in particular if severe difficulties arise in the supply of certain products, notably in the area of energy" (see under chapter VIII.a.).

A further reference to the 'spirit of solidarity' can be located in Article 194 TFEU (Title XXI, 'Energy'). This provision establishes that "[i]n the context of the establishment and functioning of the internal market and with regard for the need to preserve and improve the environment, Union policy on energy shall aim, *in a spirit of solidarity* [emphasis added] between Member States, to:

---

6 Steve Peers, 'EU Justice and Home Affairs Law' (Oxford: Oxford University Press [2011]), 311.

(a) ensure the functioning of the energy market;
(b) ensure security of energy supply in the Union;
(c) promote energy efficiency and energy saving and the development of new and renewable forms of energy; and
(d) promote the interconnection of energy networks."

The Charter of Fundamental Rights of the European Union (EUFRC) actually dedicates a whole chapter (Chapter IV, Articles 27–38) to 'Solidarity'. Title IV concerns solidarity rights, such as the right of collective action and the right to health care.

In the case 39/72 ('Premiums for slaughtering cows') the Court of Justice of the European Union (CJEU) held that the "failure in the duty of solidarity accepted by the Member States by the fact of their adherence to the Community strikes at the fundamental basis of the Community legal order"[7]. Although the founding Treaties of the European Communities did not explicitly allude to the expression 'solidarity', it is inherent to European integration. However, it is probably only since the start of the financial and economic crisis in the Eurozone that the term 'solidarity' has come to the fore of political debates[8]. With this in mind, it is noteworthy that both the Mutual Assistance Clause and the Solidarity Clause have not yet been the objects of particular interest for European analysts and even less for European politics. As mentioned above, this may, however, change in the near future as the European Council dealt with CSDP in December 2013[9]. As the 'Joint Proposal on the Implementation by the Union of the Solidarity Clause' is to be adopted by the General Affairs Council in June 2014, the Solidarity Clause may soon be placed on the agenda of European politics, too.

Despite the fact that the Mutual Assistance Clause and the Solidarity Clause appear to be closely linked through the principle of solidarity, they are divided

---

7 Commission v Italy (Case 39/72) [1973] ECR 101.

8 See Sofia Fernandes, Eulalia Rubio, 'Solidarity within the Eurozone: how much, what for, for how long?' (Notre Europe Policy Paper No. 51 [2012]), available at http://www.notre-europe.eu/media/SolidarityEMU_S.Fernandes-E.Rubio_NE_Feb2012.pdf.

9 Cf. Conclusions of the European Council (19/20 December 2013) (EUCO 217/13).

by their inclusion in two different treaties, the TEU and the TFEU, which are, however, on the same level from a legal point of view. The question therefore arises whether the scope of the two clauses is also strictly separated or whether overlaps exist or (could) come up in the implementation of the clauses. The two provisions represent two different approaches towards solidarity[10]: on the one hand, a traditional approach taken from NATO and WEU focusing on a response to a military attack on the territory of a Member State and, on the other hand, a new, more integrated approach which takes into account the 'new' threats as identified by the European Security Strategy[11]. Nevertheless, we should bear in mind that Article 222 TFEU explicitly refers the CFSP (CSDP) in its third paragraph ("[t]he Council shall act in accordance with Article 31(1) of the Treaty on European Union where this decision has defence implications [...]") whereas no links to the 'communitarised' policy areas exist in Article 42(7) TEU.

Before tackling the different questions on the relationship between the two clauses, let us have a closer look at the genesis of these two provisions.

10 See also Dolores Rubio García, ‚Las cláusulas de asistencia mutual y solidaridad introducidas por el Tratado de Lisboa: el refuerzo de la seguridad y la defensa en la Unión Europea' (Documento de Trabajo 57/2011), 20.

11 'A secure Europe in a better World - European Security Strategy', 12 December 2003. The European security strategy was drawn up under the authority of the EU's High Representative for the Common Foreign and Security Policy, Javier Solana, and adopted by the Brussels European Council of 12 and 13 December 2003. It identifies the global challenges and key threats to the security of the Union.

II. Genesis

It was the Single European Act (SEA)[12] which first explicitly referred to intra-European solidarity linked to the foreign policy area. The Preamble of the SEA asks the Member States to speak "with one voice" and "to act with consistency and solidarity":

> "AWARE of the responsibility incumbent upon Europe to aim at speaking ever increasingly with one voice and to act with consistency and solidarity in order more effectively to protect its common interests and independence, in particular to display the principles of democracy and compliance with the law and with human rights to which they are attached, so that together they may make their own contribution to the preservation of international peace and security in accordance with the undertaking entered into by them within the framework of the United Nations Charter, [...]."[13]

The Treaty of Maastricht introduced the three-pillar structure of the Union. The Common Foreign and Security Policy was established as the second (intergovernmental) pillar of the EU. Foreign policy had always been and still is considered a bastion of national sovereignty and, therefore, intergovernmentalism was and – considering the current legal framework under the Lisbon Treaty – still is considered to be better suited to safeguarding national interests within this EU policy area.[14] The Maastricht Treaty also further developed the concept of solidarity within the framework of foreign policy. Article J.1(4) asks the Member States to "support the Union's external and security policy actively and unreservedly in a spirit of loyalty and mutual *solidarity* [emphasis added]. They shall refrain from any action which is contrary to the interests of the Union or likely to impair its effectiveness as a cohesive force in international relations. The Council shall ensure that these principles are complied with."

---

12 'Single European Act', UNTS I-30614.

13 'Single European Act', 2, available at http://ec.europa.eu/economy_finance/emu_history/documents/treaties/singleeuropeanact.pdf.

14 On the development of the notion of solidarity in the CFSP cf Laura C. Ferreira-Pereira and A. J. R. Groom 'Mutual solidarity' within the EU common foreign and security policy: 'What is the name of the game?' (International Politics 47 [2010]), 596–616.

For the Treaty of Maastricht the establishment of a common defence policy was not meant to be a 'must have' but only a possible goal[15]. Therefore, Article B provides that the Union is to "assert its identity on the international scene, in particular through the implementation of a common foreign and security policy including the eventual framing of a common defence policy, which might in time lead to a common defence". In line with this, the "mutual solidarity" mentioned in Article J.1 should not establish a genuinely EU-driven collective defence obligation as according to J.4(4) "[t]he policy of the Union [...] shall not prejudice the specific character of the security and defence policy of certain Member States and shall respect the obligations of certain Member States under the North Atlantic Treaty and be compatible with the common security and defence policy established within that framework".

Article J.4(4) clearly states that the defence policy of Member States which are also NATO members continues to be covered by the NATO Treaty. Neutral Member States (at that time only Ireland) were guaranteed that the development of the defence policy of the Union should not violate the "specific character of the security and defence policy of certain Member States". This guarantee became known as the 'Irish Clause' and is also included in the current Mutual Assistance Clause. The Irish Clause exempts neutral or non-aligned countries from the obligation to assist other Member States if mutual assistance situation arises (see IV.d.).

The Treaty of Amsterdam of 1997 did not substantially overcome one of the main obstacles that hamper the development of an integrated and operational European foreign policy: unanimity voting in the Council. Although the issue of introducing qualified majority voting was intensively debated at the beginning of the 1990s and again within the framework of the Intergovernmental Conference (IGC) in 1996/1997, foreign policy was considered too sensitive an issue to transfer full sovereignty to the European

---

15 See Eduardo Gianfrancesco, 'La Politica Estera e di Sicurezza Europea prima e dopo il Trattato di Lisbona' (Forum di Quaderni Costituzionali [2011]), 3, available at http://www.forumcostituzionale.it/site/images/stories/pdf/documenti_forum/paper/0267_gianfrancesco.pdf.

Union[16]. However, the Amsterdam Treaty developed a further, very peculiar aspect of solidarity within the CFSP by introducing so-called 'constructive abstention'. The instrument of constructive abstention allows Member States not to participate in a particular measure while not hindering EU measures if the other Member States or a group of Member States has decided to agree on them. According to Article 23 TEU (Amsterdam version), "[a]bstentions by members present in person or represented shall not prevent the adoption of such decisions. When abstaining in a vote, any member of the Council may qualify its abstention by making a formal declaration under the present subparagraph. In that case, it shall not be obliged to apply the decision, but shall accept that the decision commits the Union. In *a spirit of mutual solidarity* [emphasis added], the Member State concerned shall refrain from any action likely to conflict with or impede Union action based on that decision and the other Member States shall respect its position. If the members of the Council qualifying their abstention in this way represent more than one third of the votes weighted in accordance with Article 205(2) of the Treaty establishing the European Community, the decision shall not be adopted."

The 'Irish Clause', 'constructive abstention' and obviously the instrument of 'enhanced cooperation'[17], which was also introduced by the Treaty of Amsterdam, contributed to the increasing fragmentation and differentiated

16 See Sophie Vanhoonacker, 'Common Foreign and Security Policy: Can History be overcome?' (Paper Prepared for ECSA Conference, Seattle [May 1997]), 3, available at http://aei.pitt.edu/6838/1/002722.PDF.

17 Enhanced cooperation is a procedure that allows (at least nine) Member States wishing to continue to work more closely together to do so, while respecting the legal framework of the Union. The Member States concerned can thus move forward at different speeds and/or towards different goals. However, enhanced cooperation does not allow extension of the powers as laid down by the Treaties, nor may it be applied to areas that fall within the exclusive competence of the Union. Moreover it may be undertaken only as a last resort, when it has been established within the Council that the objectives of such cooperation cannot be attained within a reasonable period by the Union as a whole (see the 'Glossary – enhanced cooperation', available at http://europa.eu/legislation_summaries/glossary/enhanced_cooperation_en.htm).

integration of the European Union's legal framework. Of course, taking into account the fact that the European Union was built upon States with different foreign policy and defence cultures (NATO, WEU, neutral and non-aligned countries) this fragmentation was clearly inevitable, at least as regards the CFSP. The Treaty of Nice, which entered into force on 1 February 2003, maintained the CFSP structure established by the Treaty of Maastricht.

II.a. Working Group VIII 'Defence' of the European Convention

The origins of the current Mutual Assistance Clause and, in particular, the Solidarity Clause can be traced to the work of the European Convention Working Group VIII on defence (2002–2003). Before analysing the results of this working group let us briefly recall the mandate of the European Convention. In their meeting of 14/15 December 2001 in Laeken the heads of State and government attributed great significance to mutual solidarity and affirmed that "[t]he European Union is a success story. For over half a century now, Europe has been at peace. Along with North America and Japan, the Union forms one of the three most prosperous parts of the world. As a result of mutual solidarity and fair distribution of the benefits of economic development, moreover, the standard of living in the Union's weaker regions has increased enormously and they have made good much of the disadvantage they were at."[18] The European Council assumed that the citizens of the European Union expected a "greater and better coordinated action to deal with trouble spots in and around Europe and in the rest of the world"[19] and posed questions as to whether "a more coherent common foreign policy and defence policy"[20] should be developed, as well as whether the 'Petersberg tasks' (humanitarian and rescue tasks, peacekeeping tasks, tasks of combat forces in crisis management, including peacemaking) should be updated.

---

18 'Presidency Conclusions - European Council Meeting 14 and 15 December 2001 in Laeken' (SN 300/1/01), 19.

19 Ibid., 21.

20 Ibid., 22.

The Working Group on Defence, which was chaired by Mr. Michel Barnier, conducted its work on the basis of its mandate (CONV 206/02 and CONV 246/02), and on the basis of questions which were put to it[21]. The Working Group was, amongst other things, asked to tackle the question of "whether an undertaking of collective defence should be enshrined in the Treaty or in a Protocol annexed thereto, possibly with an "opting-in" clause for States which might not wish to subscribe to such an undertaking as of now or which might not yet have the capabilities"[22]. Moreover, a note of Mr. Barnier's states that "following the attack on 11 September, the question arose as to whether certain actions, and if so which ones, could be undertaken by the other Member States under the Treaty in the event of a similar attack or a biological or chemical attack against a Member State"[23]. This could be interpreted as a request to the working group to consider alternative provisions to the 'classic' mutual defence approach. However, neither the original mandate (CONV 206/02) nor the note of Mr. Barnier's contain express references to a 'solidarity clause'.

According to a contribution to the Group's draft final report by a member of Working Group VIII "[t]he possible introduction of a so-called "solidarity clause" in the new Treaty was a subject of lengthy debate in the Group"[24]. However, the draft report did not fully reflect the debates held on the 'solidarity clause' nor on other issues such as the relations between the EU and NATO, crisis management and institutional arrangements[25]. Another member of the Group submitted some remarks on the need for the EU to find "answers beyond the traditional military methods"[26]. A wide concept of 'security' within the framework of the CFSP was referred to by a further member of the Working Group VIII:

---

21 See 'Final report of Working Group VIII – Defence' (CONV 461/02), 3.

22 CONV 246/02, 3. CONV 246/02 is a note drawn up by Mr. M. Barnier, Chairman of the Working Group on Defence, detailing the mandate set out in CONV 206/02 in order to facilitate the Group's discussions.

23 Ibid., 2.

24 'Comments by Mr Manuel Lobo Antunes on the preliminary draft final report of Working Group VIII "Defence" (WD 022)', 4.

25 See ibid., 2.

26 'Comments by Mrs Maria Eduarda Azevedo on the preliminary draft final report of Working Group VIII "Defence" (WD 022)', 4.

"An EU Common Foreign and Security Policy (CFSP), must be based on pro-active conflict prevention by peaceful means, in full cooperation with the UN, and fully accountable to the European Parliament and National Parliaments.

This should be founded on the twin beliefs that:

1. The international settlement of disputes should rest solely with the United Nations and must be based on the UN Charter and principles of international law.
2. Security must be broadly defined, with conflict prevention and resolution measures being directed at the full range of social, economic, political and environmental issues that breed insecurity, injustice, and war."[27]

The same member of Working Group VIII regrets that the preliminary draft report "is not at all clear" when referring "again to 'the new threat' of global terrorism" and "how a common defence/solidarity clause or increased military spending could counteract this threat"[28]. Indeed, when reading the relevant sections (31, 32 and 41) that Mr. Gormley, an Irish Group member, refers to, no clear distinction between the 'mutual defence clause' or 'mutual assistance clause' and a new 'solidarity clause' can be found. According to the preliminary draft report, the terrorist attacks of 11 September 2001 "prompt consideration not only of the need to project stability outside the Union, but also of the need to ensure security within the European Union, particularly as regards protection of the civilian population and democratic institutions"[29]. This reference to the protection of civilian population and democratic institutions will later be included in Article I-42 'Solidarity Clause' of the 'Draft Treaty Establishing a Constitution for Europe' (CONV 850/03). However, the inclusion of "an all-embracing proposal for the mobilisation of EU instruments to fight terrorism"[30] did not meet with the approval of Mr. Gormley: "This represents quite an extension of the EU's competence. A solidarity clause in the Constitution

---

27 'Comments by Mr John Gormley on the preliminary draft final report of Working Group VIII "Defence" (WD 022)', 2.

28 Ibid., 5.

29 'Preliminary draft report from the Working Group on Defence (WD 022)', 10.

30 'Comments by Mr John Gormley', 5.

which amounted to a common defence would be unacceptable. Deciding who the enemy is in a 'non-state/terrorist attack' can be extremely difficult. It is important that the threat of terrorism is not used as a pretext for the erosion of civil liberties or increased militarisation of the EU. Of course, solidarity in terms of humanitarian assistance is absolutely vital. This is a matter on which there would be broad agreement. However, the current treaty arrangements adequately cover defence matters."[31]

The relevant section of the preliminary draft report of Working Group VIII says that "[i]n application of the principle of solidarity between Member States which might be embodied in one of the initial articles of the Constitutional Treaty, a new clause could be inserted into the Treaty enabling all the instruments available to the Union to be mobilised, including the military resources and the structures originally set up for the Petersberg missions, in order to avert the terrorist threat, protect the civilian population and democratic institutions and to assist a Member State within its territory in dealing with the consequences of a possible attack"[32]. The preliminary draft further outlines the scope of the clause: "Such a clause would not be a clause on the defence of territorial integrity, but would apply to threats from non-State entities; moreover, assistance in managing the consequences of an attack would be provided only at the request of the civilian authorities of the country concerned."[33]

Mr. Gormley's comments were not the only ones that criticised the provisions on the solidarity clause in the preliminary draft. Mr. Kvist, a Swedish member of the Group, submitted the following comment:

> "P. 42 introduces the possibility of a 'solidarity clause', a kind of 'soft mutual defence commitment' for terrorist and other similar threats. Military means are presupposed to be available to avert such threats. It could be noted that in Sweden (as is the case in other Member States as well, I believe) military means cannot be used to solve internal security problems. This is a duty for the police. Paragraph 42 thereby proposes collective action going well beyond what we are entitled to do even

31 Ibid.

32 'Preliminary draft report', 14.

33 Ibid.

> individually within our own territory. This is extremely controversial and will create far more problems than it is aiming at solving. I do not see how it fits in the mandate of the Group. The paragraph should be deleted."[34]

Franco-German comments were also submitted, aiming at an amendment of the wording of the preliminary draft report. The Franco-German remarks called for, amongst other things, the solidarity clause not to be a "clause of common defence"[35]. Other members of the Working Group VIII, such as the Finnish member Mr. Seppänen, strongly opposed such a 'solidarity clause' and argued against the militarization of the EU: "I am also against the idea that the military should be involved in terrorist actions, as would be the result of such a clause"[36]. Also according to Ms. Hubner, the Polish Group member, the solidarity clause remained controversial as it would "bear the notion of collective defence and therefore should be further discussed"[37]. Moreover, "[t]he distinction between state and non-state terrorism is not helpful. Nor is it applicable in this case"[38]. The Group members Ms. Kaufmann[39] and Ms. Stuart[40] were also very critical in their comments with regard to the inclusion of a solidarity clause and the use of military means to tackle terrorism and to secure internal security of the Union.

---

34 'Comments by Mr Kenneth Kvist on the preliminary draft final report of Working Group VIII "Defence" (WD 022)', 4.

35 'Franco-German comments on the preliminary draft final report of Working Group VIII "Defence" (WD 022)', 16.

36 'Comments by Mr Esko Seppänen on the preliminary draft final report of Working Group VIII "Defence" (WD 022)', 2.

37 'Comments by Ms Danuta Hubner on the preliminary draft final report of Working Group VIII "Defence" (WD 022)', 5.

38 Ibid.

39 See 'Comments by Ms Sylvia-Yvonne Kaufmann on the preliminary draft final report of Working Group VIII "Defence" (WD 022)', 6–7.

40 See 'Comments by Ms Gisela Stuart to the preliminary draft final report of Working Group VIII "Defence" (WD 022)', 3.

On the whole positive comments on the section regarding the solidarity clause of the preliminary draft were submitted by the Group members Mr. Gaber[41] and Mr. Kiljunen[42], as well as from the Netherlands[43].

The final report of Working Group VIII, published on 16 December 2002, took into account some of the comments submitted on the preliminary draft. However, it did not substantially change its content. The final report contains in its second part the Working Group's recommendations, "several of which have received wide support"[44]. According to the final report the inclusion of a "new clause"[45] also received "broad support"[46].

"That clause would enable all the instruments available to the Union to be mobilised (including the military resources and the structures originally set up for the Petersberg tasks, as well as police and judicial cooperation, civil protection, etc.) in actions undertaken within the territory of the Union aimed, in particular, at averting the terrorist threat, protecting the civilian population and democratic institutions and assisting a Member State within its territory in dealing with the implications of a possible terrorist attack. It would therefore be a question of taking advantage of the interdisciplinary character of the Union's approach, in order both to respond effectively

---

41 'Comments by Mr Slavko Gaber on the preliminary draft final report of Working Group VIII "Defence" (WD 022)', 3.

42 'Comments by Mr Kimmo Kiljunen to the preliminary draft final report of Working Group VIII "Defence" (WD 022)', 5. This member of Working Group VIII suggested the inclusion of a reference to the Civil Protection Mechanism which was first introduced in 2001: "Efforts already undertaken within the Union, such as the creation of Civil Protection Mechanism in October 2001, should be taken into account in this work in order to avoid duplication." A focus on civil protection was also suggested by Mr. Piks in his comments to the preliminary draft: "Specialization of the Union in coordination with other institutions in "uncovered" areas would be of highest importance, for example, in civil protection area" ('Comments by Mr Rihards Piks to the preliminary draft final report of Working Group VIII "Defence" (WD 022)', 2).

43 'Dutch comments on the preliminary draft final report of Working Group VIII "Defence" (WD 022)', 4.

44 'Final report of Working Group VIII – Defence' (CONV 461/02), 1.

45 Ibid., 20.

46 Ibid.

to new challenges and to indicate clearly what distinguishes the European Union from a military alliance."[47]

While the preliminary draft stated that the "clause would not be a clause on the defence of territorial integrity"[48], the final report tried to clarify the distinction with regard to a "collective defence clause"[49], which should also be included in the 'Constitutional Treaty', and noted "that such a clause would not be a clause on collective defence entailing an obligation to provide military assistance, but would apply to threats from non-State entities"[50].

The Final Report of Working Group VIII also mentions that "[s]ome members of the Group proposed that the notion of solidarity be reflected in a broader clause on solidarity and common security which would be incorporated in the Constitutional Treaty and to which would be linked, in an annex to the Treaty, a political declaration on solidarity and common security in order to identify risks of any sort that threaten the Union, including terrorism, and the means of dealing with them"[51].

Unlike the preliminary draft, the final report includes a reference to a 'collective defence clause' but, at the same time, highlights the divergent views within the group on the inclusion of such a clause. However, the report does not outline the features of this clause and only stresses the fact that "those members of the Group who were in favour of the collective defence clause thought that it would be sensible to allow those Member States wishing to intensify their cooperation, and in particular to take over the commitments of the WEU Treaty, to do so within the framework of the Union rather than outside the Union"[52]. Therefore, the "new Treaty could establish a closer type of cooperation on defence open to all Member States wishing to enter into such a commitment and fulfilling the requirements for such a commitment to be credible, in particular in terms of command and

47 Ibid.

48 'Preliminary draft report', 14.

49 See chapter II.c) of the 'Final report', 21.

50 'Final report', 20.

51 Ibid., 21.

52 Ibid.

control capabilities, force preparedness, interoperability and deployment capabilities"[53].

All in all, the final report remains rather vague on both clauses, in particular on the Mutual Assistance Clause. The report also lacks further explanations with regard to possible mechanisms for implementing the solidarity clause. Moreover, it is notable that, although several members expressed their reluctance or even strong opposition to the inclusion of a solidarity clause, the Working Group VIII maintained the clear recommendation to provide for such a clause in the future Constitution for Europe. Finally, no explicit mention is made on other events that may trigger the clause such as natural or man-made disasters.

II.b. Provisions in the Constitution for Europe

The Treaty establishing a Constitution for Europe (TCE) built upon the work of the 'Convention on the Future of Europe' (hereinafter: Convention). The Convention agreed on the 'Draft Treaty Establishing a Constitution for Europe' in June 2003 and submitted it to the European Council. As the Member States in the European Council of Laeken did not entirely give up the normal reins of power, an intergovernmental conference (IGC) was convened in order to discuss the Draft Treaty adopted by the Convention.[54] The IGC amended some provisions in the Draft Treaty but kept the provisions on solidarity and mutual assistance. The Mutual Assistance Clause in the Draft Treaty, however, included detailed arrangements for participation in and the operation of closer cooperation on mutual defence as provided for in Article I-40(7) of the Draft. The relevant Article (III-214 of the Draft Treaty) was deleted from the final text of the TCE as the provision "remained quite unacceptable both to neutrals, who feared involvement despite the apparent opt-in mechanism, and to Atlanticists, still suspicious about the implications for NATO"[55].

---

53 Ibid., 21–22.

54 See Paul Craig, Gráienne de Búrca, 'EU Law – Text, Cases and Materials' (Oxford: Oxford University Press [2011]), 23.

55 Jolyon Howorth, 'The European Draft Constitutional Treaty and the Future of the European Defence Initiative: A Question of Flexibility' (European Foreign Affairs Review 9 [2004]), 13, available at http://www.cap.lmu.de/transatlantic/download/howorth.pdf.

II.b.i. The Mutual Assistance Clause

Before we go into the provisions of the Treaty establishing a Constitution for Europe (TCE), let us recall the so-called Tervuren meeting of the Heads of State and Government of Germany, France, Luxembourg and Belgium. In their declaration on European defence of 29 April 2003 these countries asked for the inclusion of a "general clause of solidarity and common security":

> "In order to give new impetus to the European Security and Defence Policy, we propose that the Convention on the Future of Europe and the Intergovernmental Conference approve the following principles and integrate them into the constitutional Treaty:
>
> […] A general clause of solidarity and common security binding all member States of the European Union and making it possible to face up to all kinds of risks concerning the European Union."[56]

The Final Report of Working Group VIII had already given rise to a group of Member States asking for "a broader clause on solidarity and common security"[57] but still the draft 'Constitutional Treaty'[58] as adopted by consensus by the European Convention on 13 June and 10 July 2003[59] limited binding mutual assistance in cases of armed aggression to those Member States who have established closer cooperation in the field of defence[60]. The ministers of

56 'Joint Statement of the Heads of State and Government of Germany, France, Luxembourg and Belgium on European Defence, Brussels, 29 April 2003' ("Tervuren Declaration"), available at http://www.ambafrance-uk.org/Meeting-of-the-Heads-of-State-and.html.

57 'Final report', 21.

58 'Draft Treaty establishing a Constitution for Europe' (CONV 850/03).

59 Official Journal of the European Union C 169/3, 'Draft Treaty establishing a Constitution for Europe', 18.7.2003.

60 See Article I-40(7) of the Draft Treaty establishing a Constitution for Europe: "Until such time as the European Council has acted in accordance with paragraph 2 of this Article, closer cooperation shall be established, in the Union framework, as regards mutual defence. Under this cooperation, if one of the Member States participating in

foreign affairs of the neutral States (Austria, Finland, Ireland and Sweden) underlined their reluctance to include a collective defence clause automatically binding all Member States in a letter to the then President of the Council, Mr. Frattini[61].

Finally, despite the position of the neutral States outlined above, the TCE provided for a clause binding all Member States which was included in Part I, Title V (Exercise of Union Competence), Chapter II (Specific Provisions), Article I-41 (7):

---

such cooperation is the victim of armed aggression on its territory, the other participating States shall give it aid and assistance by all the means in their power, military or other, in accordance with Article 51 of the United Nations Charter." This provision is further elaborated in Article III-214(2): "A Member State participating in such cooperation which is the victim of armed aggression on its territory shall inform the other participating States of the situation and may request aid and assistance from them. Participating Member States shall meet at ministerial level, assisted by their representatives on the Political and Security Committee and the Military Committee." Therefore, the Draft Treaty provided for the prior involvement of an EU institution, a provision that was later left out in the TCE. Article III-214(2) can be seen as an element of a centralised collective security system where the decision with regard to a situation of collective defence is taken by an organ of the collective security system. Unlike a centralised collective security system, a decentralised system confers the final decision of whether a situation triggering the collective defence applies on the individual Member State of the system. The latter solution is reflected in the TCE and the Lisbon Treaty (see Gerhard Hafner, ‚Beistands- und Solidaritätsklausel', in Waldemar Hummer, Walter Obwexer, ‚Der Vertrag von Lissabon' (Baden-Baden: Nomos Verlag [2009]), 385).

61 "We would also like to stress the importance to enhance mutual solidarity among EU Member States. We fully support the proposal to add in the Constitutional Treaty a solidarity clause which would take into account terrorism and natural and man-made disasters. Moreover, we are prepared to underline the principle of EU solidarity more widely in the field of security, including in situations referred to in Article 51 of the UN Charter. However, provisions containing formal binding security guarantees would be inconsistent with our security policy or with our constitutional requirements. We therefore propose the following text for the first paragraph of Article 40(7) of the Constitutional Treaty: If a Member State is victim of armed aggression, it may request that the other Member States give it aid and assistance by all the means in their power, military or other, in accordance with article 51 of the UN Charter." (CIG 62/03, 5 December 2003, 2.)

> "If a Member State is the victim of armed aggression on its territory, the other Member States shall have towards it an obligation of aid and assistance by all the means in their power, in accordance with Article 51 of the United Nations Charter. This shall not prejudice the specific character of the security and defence policy of certain Member States Commitments and cooperation in this area shall be consistent with commitments under the North Atlantic Treaty Organisation, which, for those States which are members of it, remains the foundation of their collective defence and the forum for its implementation."

This formula which, on the one hand, contains a binding collective defence commitment but, on the other hand, qualifies this commitment by introducing a reference to the "specific character" of the security and defence policy of the neutral Member States, was later kept in the Lisbon Treaty[62].

II.b.ii. The Solidarity Clause

In the note on the 'Draft Articles on external action in the Constitutional Treaty' the Praesidium of the Convention included the following commentary on the provision referring to the Solidarity Clause:

> "In view of the fundamental and constitutional nature of the clause, it appears in Part One of the Constitution. However, it should be noted that it is of horizontal scope, enabling the various resources to be called on (both national military capabilities and the instruments of the Union). Consequently, it is covered by a separate article in Title V of Part I of the Constitution."[63]

Indeed, the Solidarity Clause was finally introduced by the European Convention in two parts: Article I-42 of the draft Constitution was named 'Solidarity Clause' and was situated in Chapter II (Specific provisions) under Title V (Exercise of Union competence). Article III-231 came under the Chapter VIII headline 'Implementation of the solidarity clause'. The TCE maintained this

---

62 See Hafner, ‚Beistands- und Solidaritätsklausel', 385.

63 CONV 685/03, 22.

structure and, hence, we find the corresponding provisions in the Constitutional Treaty in Articles I-43 and III-329. Therefore, unlike the current provision introduced by the Lisbon Treaty, we have a division into two parts. In Article I-43 we find the core elements of the Union's obligation:

"1. The Union and its Member States shall act jointly in a spirit of solidarity if a Member State is the object of a terrorist attack or the victim of a natural or man-made disaster. The Union shall mobilise all the instruments at its disposal, including the military resources made available by the Member States, to:
(a) — prevent the terrorist threat in the territory of the Member States;
— protect democratic institutions and the civilian population from any terrorist attack;
— assist a Member State in its territory, at the request of its political authorities, in the event of a terrorist attack;
(b) assist a Member State in its territory, at the request of its political authorities, in the event of a natural or man-made disaster.
2. The detailed arrangements for implementing this Article are set out in Article III-329."

Whereas Article I-43 outlines the 'substantive' elements of the clause, Article III-329 comprises the implementing provisions:

"1. Should a Member State be the object of a terrorist attack or the victim of a natural or man-made disaster, the other Member States shall assist it at the request of its political authorities. To that end, the Member States shall coordinate between themselves in the Council.
2. The arrangements for the implementation by the Union of the solidarity clause referred to in Article I-43 shall be defined by a European decision adopted by the Council acting on a joint proposal by the Commission and the Union Minister for Foreign Affairs. The Council shall act in accordance with Article III-300(1) where this decision has defence implications. The European Parliament shall be informed.
For the purposes of this paragraph and without prejudice to Article III-344, the Council shall be assisted by the Political and Security

Committee with the support of the structures developed in the context of the common security and defence policy and by the Committee referred to in Article III-261; the two committees shall, if necessary, submit joint opinions.

3. The European Council shall regularly assess the threats facing the Union in order to enable the Union and its Member States to take effective action."

Therefore, the obligation of the Member States to assist another Member State in the case of the latter becoming the object of a terrorist attack, a natural or a man-made disaster is enshrined in paragraph (1) of the implementing provision. At first glance, it appears to be as if it is first and foremost the duty of the Member States to implement the Solidarity Clause. However, this indication is overruled by paragraph (2) which calls for a further (secondary law) implementing provision, this time referring to the arrangements for the implementation by the Union of the Clause. The original structure of the Solidarity Clause in the TCE will in any case remain relevant for the interpretation of Article 222 TFEU (see chapter V.).

II.c. European Council conclusions 2004

Following the terrorist attacks in Madrid on 11 March 2004, the heads of State or government of the Member States of the European Union, and of the States acceding to the Union on 1 May 2004, declared their firm intention to act in the spirit of the solidarity clause laid down in Article 42 of the draft Treaty establishing a Constitution for Europe, and to act jointly in a spirit of solidarity if one of them is the victim of a terrorist attack, by mobilising all the instruments at their disposal, including military resources[64]. In an annex to the declaration the heads of State or government not only reiterated their "firm intention" to act in a spirit of solidarity but also set limits to this obligation of the Member States by stating that it is to be an autonomous decision "for each Member

64 European Council, 'Declaration on solidarity against terrorism', 25 March 2004.

State or acceding State to the Union to choose the most appropriate means to comply with this solidarity commitment towards the affected State"[65]. Hence, this reference to the right to choose "the appropriate" means anticipated the meaning of Declaration No. 37 to Article 222 TFEU (cf. chapter V.a.ii.).

The 11-M bombings (as they are usually called in Spain) in Madrid and the terrorist attack in London more than a year later (7 July 2005) paved the way to the adoption of an EU strategy against terrorism[66] – which was deeply influenced by the UK strategy against terrorism[67] – including an action plan as well as a strategy against radicalisation and recruitment[68]. These strategic papers related to the fight against terrorism and the declaration of 25 March 2004 may have indeed "embodied a collective EU solidarity statement against terrorism and a 'call to arms' addressed to the Member States' national authorities being the primary resources to carry out such a course of action with the assistance

65 'Draft Declaration on Solidarity against Terrorism' (contained in the Annex to the Declaration of 25 March 2004):
"We, the Heads of State or Government of the Member States of the European Union, and of the States acceding to the Union on 1 May, have declared our firm intention as follows: In the spirit of the solidarity clause laid down in Article 42 of the draft Treaty establishing a Constitution for Europe, the Member States and the acceding States shall accordingly act jointly in a spirit of solidarity if one of them is the victim of a terrorist attack. They shall mobilise all the instruments at their disposal, including military resources to:
- prevent the terrorist threat in the territory of one of them;
- protect democratic institutions and the civilian population from any terrorist attack;
- assist a Member State or an acceding State in its territory at the request of its political authorities in the event of a terrorist attack.

It shall be for each Member State or acceding State to the Union to choose the most appropriate means to comply with this solidarity commitment towards the affected State."

66 'The European Union Counter-Terrorism Strategy' (14469/4/05 Rev 4).

67 Cf. the revised version 'The United Kingdom's Strategy for Countering Terrorism' published in July 2011 which keeps its typical structure (Pursue, Prevent, Protect, Prepare), available at https://www.gov.uk/government/uploads/system/uploads/attachment_data/file/97994/contest-summary.pdf.

68 'The European Union Strategy for Combating Radicalisation and Recruitment to Terrorism' (14781/1/05).

of EU bodies such as Europol and Eurojust"[69]. Be that as it may, the tragic events of 9/11, Madrid and London made the commitment of the Union to fostering solidarity in the event of major terrorist attacks irrevocable.

69 Theodore Konstadinides, 'Civil Protection in Europe and the Lisbon 'solidarity clause': A genuine legal concept or a paper exercise' (Uppsala Faculty of Law, Working Paper [2011]), available at http://www2.statsvet.uu.se/LinkClick.aspx?fileticket=ywEQQ722UuI%3D&tabid=3159&language=sv-SE, 11.

III. Models of the Mutual Assistance Clause

Article 42(7) of the Treaty on European Union provides that "[i]f a Member State is the victim of armed aggression on its territory, the other Member States shall have towards it an obligation of aid and assistance by all the means in their power, in accordance with Article 51 of the United Nations Charter. This shall not prejudice the specific character of the security and defence policy of certain Member States. Commitments and cooperation in this area shall be consistent with commitments under the North Atlantic Treaty Organisation, which, for those States which are members of it, remains the foundation of their collective defence and the forum for its implementation". The Mutual Assistance Clause has models such as the NATO treaty and the modified Brussels Treaty of 1954, which established the Western European Union (WEU), whereas the Solidarity Clause is a genuine innovation of the European Union.

III.a. Article V Western European Union

It is commonly assumed that Article 42(7) TEU has its origins in Article V of the modified Brussels Treaty of 1954, which established the Western European Union (WEU). Article V Brussels Treaty reads as follows:

> "If any of the High Contracting Parties should be the object of an armed attack in Europe, the other High Contracting Parties will, in accordance with the provisions of Article 51 of the Charter of the United Nations, afford the Party so attacked all the military and other aid and assistance in their power."

This commitment to mutual defence should any of its signatories become the victim of an armed attack in Europe constituted the cornerstone of the now defunct Western European Union[70], an institution dedicated to collective

70 WEU members were Belgium, France, Germany, Greece, Italy, Luxembourg, Netherlands, Portugal, Spain, and United Kingdom.

European self-defence[71]. The territorial scope of the mutual defence clause of the Brussels Treaty is limited to Europe. However, the obligation to assist the attacked High Contracting Party is binding and comprehensive ("the other High Contracting Parties *will*[72] [...] afford [...] *all the military and other aid and assistance*[73] in their power").

III.b. Article 5 NATO Treaty

The North Atlantic Treaty (or more popularly known as 'Washington Treaty') of 1949 enshrines in its Article 5 a collective defence clause:

> "The Parties agree that an armed attack against one or more of them in Europe or North America shall be considered an attack against them all and consequently they agree that, if such an armed attack occurs, each of them, in exercise of the right of individual or collective self-defence recognised by Article 51 of the Charter of the United Nations, will assist the Party or Parties so attacked by taking forthwith, individually and in concert with the other Parties, such action as it deems necessary, including the use of armed force, to restore and maintain the security of the North Atlantic area. Any such armed attack and all measures taken as a result thereof shall immediately be reported to the Security Council. Such measures shall be terminated when the Security Council has taken the measures necessary to restore and maintain international peace and security."

The territorial scope of this clause is outlined in Article 6(1) of the NATO Treaty:

> "For the purpose of Article 5, an armed attack on one or more of the Parties is deemed to include an armed attack:

- on the territory of any of the Parties in Europe or North America, on the Algerian Departments of France (2), on the territory of or on the Islands

71 See Myrto Hatzigeorgopoulos, 'The Mutual Assistance and Solidarity Clauses' (ISIS Europe European Security Review 61 [2012]), 1.

72 Emphasis added.

73 Emphasis added.

under the jurisdiction of any of the Parties in the North Atlantic area north of the Tropic of Cancer;
- on the forces, vessels, or aircraft of any of the Parties, when in or over these territories or any other area in Europe in which occupation forces of any of the Parties were stationed on the date when the Treaty entered into force or the Mediterranean Sea or the North Atlantic area north of the Tropic of Cancer."

Unlike the WEU Treaty, the Washington Treaty leaves the decision on the extent and the means of its assistance to the assisting State ("such action as it deems necessary")[74].

---

74 See Hafner, ‚Beistands- und Solidaritätsklausel', 381.

IV. Definition and Scope of the Mutual Assistance Clause

After numerous attempts to include the mutual defence obligation stemming from the WEU in EU primary law, the Lisbon Treaty finally incorporated a mutual aid and assistance clause in its Article 42(7) TEU[75]. Due to its location within the TEU in Section 2 on the CSDP, the Mutual Assistance Clause (sometimes also called 'Mutual Defence Clause')[76] has a clear intergovernmental remit[77]. The Clause "can be described as purely intergovernmental in nature: it binds member states 'horizontally' without transferring any competence 'vertically' to EU institutions; nor does it require coordination at the EU level in situations when the mutual defence obligation is invoked. In a legal sense, Article 42.7 differs from most treaty provisions because it is only weakly associated with EU institutions and EU level capabilities."[78]

Article 42 can be considered as a further development of the old Article 17 TEU (Treaty of Nice)[79]. Of all the different provisions in Article 42, it is worth mentioning paragraph (2) which reiterates the possible future common defence commitment introduced by the Treaty of Maastricht, whereas the

75 See Hatzigeorgopoulos, 2.

76 The term 'mutual defence' is more narrow than the term 'mutual assistance' as it implies that the assistance provided by the Member States to the Member State being the victim of an armed aggression is exclusively directed towards defending (obviously by armed military means) the territory of the attacked Member State against further attacks from outside. This kind of 'defence' would probably also mean that the attacked Member State and its 'allies' may also be obliged to perform counter-attacks in order to secure the defence lines or simply to prevent further attacks. The term 'assistance' is broader as it would also encompass non-armed military assistance such as the provision of specific equipment to combat the use of chemical weapons or the provisional re-establishment of damaged infrastructure, etc.

77 See Garcia, 39.

78 Myrdal, Rhinard, 12.

79 See Hafner, ‚Beistands- und Solidaritätsklausel', 379.

Mutual Defence Clause in paragraph (7) is the directly applicable obligation to mutually assist in the event of armed aggression[80].

IV.a. Definition

The definition of the Mutual Assistance Clause in Article 42(7) TEU is obviously taken from Article V of the Brussels Treaty[81]. If we compare both provisions, we can easily find structural similarities, but also differences. Differences are particularly evident with regard to the assistance that is to be afforded to the Contracting Party or Member State[82]. As regards the similarities to Article 5 of the Treaty of Washington, these are also evident although Article 42(7) TEU does not explicitly mention 'the use of armed force'. "What is more, contrary to NATO, the EU does not maintain strong military readiness structures or joint military planning. [...] [M]utual assistance at EU level still has to be compromised with the broader security commitments of the Member States in NATO, especially those related to collective self-defence under Article 51 of the Washington Treaty 1949."[83]

Despite apparent similarities, from a practical point of view there are a "number of fundamental and overlooked differences" which "render impossible the confusion of Art. 42(7) for a mutual defence clause along the lines of NATO's Art. V or Art. V of the now-defunct modified Brussels Treaty"[84]. "Firstly, the provision's inclusion within the Lisbon Treaty automatically reaches a bigger number of recipients, NATO and non-NATO members alike. If the WEU comprised only 10 members, all members of NATO, the obligation of the Lisbon Treaty applies equally to all 27 EU Member States, including the neutral ones. The extension of the clause's obligation to all EU Member States, coupled with the direct reference to NATO as the key

---

80 Ibid., 380.

81 See Christoph Vedder, ‚Artikel I-41', in Christoph Vedder, Wolff Heintschel von Heinegg, ‚Europäischer Verfassungsvertrag' (Baden-Baden: Nomos Verlag [2007]), 209.

82 See ibid., 381.

83 Konstadinides, 18.

84 Hatzigeorgopoulos, 3.

security arrangement in Europe, creates a difficult and controversial situation where the credibility and efficiency of the clause appear to be limited. Art. 42(7), indeed, differs from Art. V of the modified Brussels Treaty in that it entails a different appreciation of the type of relation between EU and NATO; while the WEU Treaty referred to NATO as a partner for cooperation in the application and implementation of the mutual defence clause, the TEU explicitly refers to the Alliance for the implementation of mutual assistance. The Treaty enshrines the role of NATO as the foundation for collective defence of states that are members of both NATO and the EU."[85]

Returning to the definition of the Clause, it is worth considering that the Washington and Brussels Treaties, as well as Article 51 UN Charter, use the term 'armed attack' whereas Article 42(7) refers to the term 'armed aggression'. Without getting further involved in this discussion, it should be mentioned (as clearly outlined by Hafner) that no different meanings can be derived from the diverging terminology; in particular, there seems to be no indication that the apparently more narrow term 'aggression' aims to exclude terrorist attacks, which are explicitly mentioned in Article 222 TFEU[86].

IV.b. Territorial Scope

The geographical scope of the Mutual Assistance Clause refers to the 'territory' of a Member State. What is the territory of a Member State according to EU law? As stated in Article 52 TEU, the territorial scope of the Treaties is specified in Article 355 TFEU[87]. However, some Member States could extend

85 Hatzigeorgopoulos, 3–4.

86 See Hafner, ‚Beistands- und Solidaritätsklausel', 383.

87 Article 355 TFEU:
"In addition to the provisions of Article 52 of the Treaty on European Union relating to the territorial scope of the Treaties, the following provisions shall apply:
1. The provisions of the Treaties shall apply to Guadeloupe, French Guiana, Martinique, Réunion, Saint Barthélemy, Saint Martin, the Azores, Madeira and the Canary Islands in accordance with Article 349.
2. The special arrangements for association set out in Part Four shall apply to the overseas countries and territories listed in Annex II. The Treaties shall not apply to those

'their' territory beyond the provisions of Article 52 TEU and 35 TFEU and refer to the territory in accordance with their national law. It could, therefore, be that – as we will see with regard to the Solidarity Clause and its implementing decision – the territory of the Member States is even extended to critical infrastructure, airplanes or vessels. Although the premises of an embassy, airplanes or vessels cannot be considered part of the territory of the home State according to international law[88], disputes between Member States could also arise.

The geographical scope of the Mutual Assistance Clause is also wider than in the WEU and even the NATO Treaty. Article V Brussels Treaty limits the scope to Europe. Article 6 Washington Treaty gives a very precise definition

---

overseas countries and territories having special relations with the United Kingdom of Great Britain and Northern Ireland which are not included in the aforementioned list.

3. The provisions of the Treaties shall apply to the European territories for whose external relations a Member State is responsible.
4. The provisions of the Treaties shall apply to the Åland Islands in accordance with the provisions set out in Protocol 2 to the Act concerning the conditions of accession of the Republic of Austria, the Republic of Finland and the Kingdom of Sweden.
5. Notwithstanding Article 52 of the Treaty on European Union and paragraphs 1 to 4 of this Article:
   (a) the Treaties shall not apply to the Faeroe Islands;
   (b) the Treaties shall not apply to the United Kingdom Sovereign Base Areas of Akrotiri and Dhekelia in Cyprus except to the extent necessary to ensure the implementation of the arrangements set out in the Protocol on the Sovereign Base Areas of the United Kingdom of Great Britain and Northern Ireland in Cyprus annexed to the Act concerning the conditions of accession of the Czech Republic, the Republic of Estonia, the Republic of Cyprus, the Republic of Latvia, the Republic of Lithuania, the Republic of Hungary, the Republic of Malta, the Republic of Poland, the Republic of Slovenia and the Slovak Republic to the European Union and in accordance with the terms of that Protocol;
   (c) the Treaties shall apply to the Channel Islands and the Isle of Man only to the extent necessary to ensure the implementation of the arrangements for those islands set out in the Treaty concerning the accession of new Member States to the European Economic Community and to the European Atomic Energy Community signed on 25 January 1972."

88 Cf. Ian Brownlie, 'Principles of Public International Law' (Oxford: Oxford University Press [2008]), 105–122 and 369–379.

of the geographical scope of the Treaty. Therefore, Article 42(7) TEU could apply when the Article 5 NATO Treaty does not apply[89].

IV.c. Instruments

The Mutual Assistance Clause makes no mention of how EU Member States should assist one another. Member States have to provide assistance "by all the means in their power". In principle, the Clause allows for many forms of assistance. In practice, the explicit reference to armed aggression points most specifically to military means. Here again we should recall that Article V WEU Treaty explicitly refers to "all the military and other aid and assistance in their power"[90]. The WEU highlights the military means and puts them as quasi first option, whereas the NATO Treaty only asks the Party to take "such action as it deems necessary, including the use of armed force"[91]. Certainly, the notion "military means" is wider than the notion of "armed force" as it also encompasses all non-armed military means. By not mentioning any particular means of assistance the TEU (unlike the TFEU in its Solidarity Clause) underlines the broadened scope of the CSDP compared to the WEU and, hence, the variety of situations in which Member States can become active within the CSDP.

Looking at the possible causes triggering the Mutual Assistance Clause, a similar conclusion can be drawn. The EU mutual assistance provision in Article 42(7) TEU does not differentiate between types of armed aggression. The aggression can result from an inter-state conflict, a State-sponsored terrorist attack (such as 9/11)[92] or even a CBRN[93] attack. Therefore the instruments of

---

89 See Hafner, ‚Beistands- und Solidaritätsklausel', 385.

90 See Myrdal, Rhinard, 10.

91 Article 5 NATO Treaty.

92 In the Nicaragua case the ICJ used the word "aggression" in order to interpret the term "armed attack". An armed attack must not necessarily be performed by regular forces. Attacks of armed bands, irregulars or mercenaries are also covered. More relevant for the notion of armed attack should be the scope and the effect, such as the number of human casualties. The territory involved could also be taken into consideration for this kind of assessment.

93 Chemical, Biological, Radiological, and Nuclear.

assistance may also vary and must not necessarily be military means. Having said that, it is rather obvious that the inclusion of the Clause in the CFSP part of the TEU indicates that traditional military armament would be most likely the main instrument used in a case of mutual assistance. However, the meaning of armed aggression may evolve over time and include also non-traditional arms (such as Trojans or other 'cyber weapons' used for a cyber attack)[94]. Also NATO has reacted to the new challenges posed by cyber threats and has adopted a new cyber defence policy[95] and an associated Action Plan in June 2011. Therefore, cyber defence would also fall within the remit of the CSDP which may give rise to some overlaps with the Solidarity Clause.

IV.d. Status of Neutral and Non-Aligned Member States

The Mutual Assistance Clause places an obligation on the Member States. This affirmative duty has been drafted in a "relatively unambiguous"[96] way, by saying that "the other Member States shall have towards it an obligation of aid and assistance by all the means in their power"[97]. Of course, this obligation is mitigated by the second sentence of Article 42(7) which is commonly known as the 'Irish Clause': "This shall not prejudice the specific character

94 As regards the possible events triggering the Mutual Assistance and the Solidarity Clauses see also Hatzigeorgopoulos, 10.

95 NATO's cyber defence policy "sets out a clear vision of how the Alliance plans to bolster its cyber efforts. This policy reiterates that any collective defence response is subject to decisions of the North Atlantic Council, NATO's principal political decision-making body. The revised policy offers a coordinated approach to cyber defence across the Alliance. It focuses on the prevention of cyber attacks and building resilience. All NATO structures will be brought under centralised protection and NATO will enhance its capabilities to deal with the vast array of cyber threats it currently faces, including through integrating them into the NATO Defence Planning Process" (taken from http://www.nato.int/cps/en/SID-982C78A7-0BBBED98/natolive/topics_78170.htm?).

96 Michael J. Austin, 'Style or Substance? An Analysis of the Major Reforms to CFSP by the Treaty of Lisbon' (European Union Law Working Papers No. 1, Stanford/Vienna Transatlantic Technology Law Forum [2011]), 13.

97 Article 42(7), first sentence, TEU.

of the security and defence policy of certain Member States."[98] In response to fears that the Clause could undermine NATO, the Treaty also states that said solidarity should be "consistent with the commitments under NATO which, for those States that are members of it, remains the foundation of their collective defence and the forum for its implementation"[99].

There were "long-standing plans for the EU to subsume the WEU's institutions, functions and responsibilities, including its mutual defence clause"[100] which could not be concretised until the IGC 2003/2004. While the 'Petersberg tasks' had been integrated in the former Article 17 TEU, there were plans to include the mutual assistance obligation of Article V WEU Treaty in a protocol to the TEU[101]. This protocol would have allowed not only the neutral or non-aligned Member States but also the United Kingdom the possibility to choose to 'opt-into' the system of collective self-defence by ratifying the additional protocol[102]. However, this 'opting-in' solution raised a number of difficult questions and was finally dropped.

A solution was found in the so-called Irish Clause, which had been already established by the Treaty of Maastricht in its Article J.4[103]. The Irish Clause in its original setting was aimed at preserving the special character of the security policy of the sole neutral Member State of the European Union at that time with regard to the newly established Common Foreign and Security Policy. During the IGC 2003/2004 the ministers of foreign affairs of the neutral States

98 Article 42(7), second sentence, TEU.

99 Article 42(7), subparagraph, TEU.

100 Karen Devine, ,Irish Neutrality and the Lisbon Treaty', available at http://doras.dcu.ie/14898/1/Irish_Neutrality_and_the_Lisbon_Treaty.pdf, 4.

101 See Waldemar Hummer, ,Paradigmenwechsel in der „Zweiten Säule" der Europäischen Union. Von der „WEU-geführten" zur „EU-geführten" Sicherheits- und Verteidigungspolitik', in Michael Gehler, Anton Pelinka, Günter Bischof (Hg.), ,Österreich in der Europäischen Union. Bilanz seiner Mitgliedschaft' (Wien: Böhlau Verlag [2003]), 315.

102 See ibid.

103 Article J.4(4): „The policy of the Union in accordance with this Article shall not prejudice the specific character of the security and defence policy of certain Member States and shall respect the obligations of certain Member States under the North Atlantic Treaty and be compatible with the common security and defence policy established within that framework."

(Austria, Finland, Ireland and Sweden) asked for the inclusion of a non-binding commitment to mutual assistance[104]. The Treaty of Lisbon not only mitigated the mutual assistance obligation by including the Irish Clause in Article 42(7) TEU but even multiplied the references to the Irish Clause. Indeed it can be found in Article 42(2) (and, of course, Article 42(7)), in the 'Protocol (No 10) on Permanent Structured Cooperation established by Article 42 of the Treaty on European Union'[105] as well as in the declarations concerning the common foreign and security policy No. 13 and No. 14[106].

What does the Irish Clause in Article 42(7) mean in practice for the neutral or non-aligned Member States? Does it really exempt them from any obligation to assist a Member State that has been the victim of armed aggression? According to Hafner[107] the Irish Clause prevents neutral or non-aligned Member States from being subject to the principles of solidarity and loyalty. Pursuant to this view, exemption from the mutual assistance obligation has an 'asymmetrical effect': on the one hand it obliges the non-neutral Member States to provide assistance in every case falling under Art. 42(7) – even to the neutral or non-aligned Member States – but dispenses the latter from the obligation to assist[108]. Others like Rehrl refuse this 'asymmetrical effect' and emphasize the principle of loyalty (or more precisely the principle of sincere (or loyal) cooperation enshrined in Article 4(3) TEU[109] but also reflected in Article 24(3)

104 See the letter of the four Member States, CIG 62/03, 2: "If a Member State is victim of armed aggression, it may request that the other Member States give it aid and assistance by all the means in their power, military or other, in accordance with article 51 of the UN Charter."

105 "RECALLING that the common security and defence policy of the Union does not prejudice the specific character of the security and defence policy of certain Member States, […]."

106 See Hafner, ‚Beistands- und Solidaritätsklausel', 395.

107 See Hafner, ‚Beistands- und Solidaritätsklausel', 395.

108 Ibid.

109 Article 4(3) TEU:
"Pursuant to the principle of sincere cooperation, the Union and the Member States shall, in full mutual respect, assist each other in carrying out tasks which flow from the Treaties. The Member States shall take any appropriate measure, general or particular, to ensure fulfilment of the obligations arising out of the Treaties or resulting from the acts of the institutions of the Union.

TEU[110])[111]. A third view could be the one outlined by Hilpold which indicates as a possible solution scrutiny on a case-by-case basis whether the neutral or non-aligned Member States could – without breaching the specific status of their security and defence policy – provide assistance in line with the principle of solidarity. Of course, according to Hilpold, the other Member States should not expect a 'blind' (unconditional) solidarity[112].

The possible scope of the Mutual Assistance Clause was also a matter of concern for the Irish government when it negotiated further 'guarantees' as conditions for submitting the Lisbon Treaty to a second referendum after the first referendum had failed to reach a majority of affirmative votes. Therefore, on 19 June 2009, the European Council agreed on the 'guarantees' requested by the Irish government. 'These guarantees' consisted of:

a) a Decision of EU heads of State and government relating to three specific issues: the right to life, family and education; tax issues; and defence;
b) a Declaration of the European Council relating to workers' rights, social policy and public services; and
c) a declaration by Ireland relating to defence issues[113].

The Member States shall facilitate the achievement of the Union's tasks and refrain from any measure which could jeopardise the attainment of the Union's."

110 Article 24(3) TEU:
"The Member States shall support the Union's external and security policy actively and unreservedly in a spirit of loyalty and mutual solidarity and shall comply with the Union's action in this area."

111 Jochen Rehrl, ‚Beistandsgarantie und Solidaritätsklausel. Völker- und Verfassungsrechtliche Herausforderungen für Österreich durch den neuen EU-Verfassungsvertrag' (Zeitschrift für öffentliches Recht 60 [2005]), 36–38. According to Rehrl the reference to the "specific character of the security and defence policy of certain Member States" does not exempt them from the obligation to assist as the assistance would be part of the „Grundwert" (basic value) 'solidarity' which constitutes the basis of the whole primary law framework. An exemption would, therefore, be meaningless (see Rehrl, 38).

112 Peter Hilpold, ‚Österreichs Neutralität nach Lissabon im Lichte der Beistands- und Solidaritätsverpflichtung' (ÖJZ 2010/65), 593–594.

113 See Steve Peers, 'Analysis - Lisbon Treaty guarantees for Ireland, 19 June 2009', available at http://www.statewatch.org/news/2009/jun/lisbon-ireland.pdf, 1.

The ‘Protocol on the concerns of the Irish people on the Treaty of Lisbon’ introduces some new elements to the discussion by stating that “[i]t will be for Member States - including Ireland, acting in a spirit of solidarity and without prejudice to its traditional policy of military neutrality - to determine the nature of aid or assistance to be provided to a Member State which is the object of a terrorist attack or the victim of armed aggression on its territory” (Article 3)[114]. Surprisingly, the protocol seems to refer to both the Mutual Assistance Clause and the Solidarity Clause as it distinguishes between terrorist attacks and armed aggression. It also clarifies the type of the Irish neutrality, namely “military neutrality”, and by underlining the right of autonomous determination “of aid or assistance” appears to exclude the possibility of complete non-assistance of the neutral or non-aligned Member States.

The fact that the Irish Clause qualifies the obligation to mutually assist does not necessarily mean that the neutral and non-aligned Member States could be completely exempted from any obligation arising out of Article 42(7) TEU. The recently adopted protocol on the concerns of the Irish people also suggests that the option of ‘no assistance at all’ is *not* an option. It has rather to be decided on a case-by-case basis (taking into account the specific status of the security and defence policy, hence, the type of neutrality, of the relevant Member States and the particular case of armed aggression triggering the Mutual Assistance Clause) which kind of assistance could be expected from neutral and non-aligned Member States. ‘Absolute’ neutrality seems indeed difficult to combine with participation in the CFSP and CSDP. Certainly, neutral Member States cannot be obliged to perform defensive military acts, if any, which would go beyond the protection of the civilian population. However, certain demonstrations of solidarity by way of diplomatic or non-military means (except those used within a ‘civil defence’ context, such as for civil protection) should also be expected from the neutral and non-aligned Member States.

---

114 Official Journal of the European Union L 60/131, ‘Protocol on the concerns of the Irish people on the Treaty of Lisbon’, 2.3.2013.

IV.e. Legal Status of the Mutual Assistance Clause

As already mentioned, the Mutual Assistance Clause stays in the 'intergovernmental' part of the Treaties as it is included in Chapter 2, Section 2 'Provisions on the Common Security and Defence Policy' of the TEU. Therefore, in accordance with Article 24 TEU, the Court of Justice of the European Union does not have jurisdiction with respect to the provisions of the CSDP including Article 42(7) TEU. The Common Foreign and Security Policy is subject to specific rules and procedures. It is to be defined and implemented by the European Council and the Council acting unanimously, except where the Treaties provide otherwise. The adoption of legislative acts is to be excluded. The CFSP "shall be put into effect by the High Representative of the Union for Foreign Affairs and Security Policy and by Member States, in accordance with the Treaties"[115]. The Mutual Assistance Clause, therefore, lacks enforceability[116]. Although the High Representative may apply "public pressure to Member States avoiding obligations"[117] under the Clause, "the lack of jurisdiction by the European Court of Justice would appear to make it legally impossible by means of an EU institution to enforce an obligation under the mutual defence clause"[118]. Indeed the lack of enforceability is "problematic to a cohesive CFSP because it prevents a Member State from being required to perform its obligations under the clause"[119].

Despite the lack of jurisdiction of the Court of Justice of the European Union, the inclusion of a new legal obligation could represent a first important step towards a more integrated approach in the field of security and defence policy. Certainly, as has often been the case in the history of European integration, Member States will only be willing to create an integrated defence policy if public pressure is stronger than the fear of losing sovereignty. An attack from outside the Union – even if it may seem unlikely (though serious cyber

115 Article 24(1) TEU.

116 See Austin, 14.

117 Austin, 14.

118 Ibid., 14–15.

119 Ibid., 14.

attacks from outside the EU have already occurred in the recent past)[120] – could rather swiftly trigger a process of further integration of the CSDP and may increasingly call into question the specific character of the security policies of some Member States.

120 The Cyber attacks against Estonia began on 27 April 2007 and targeted the websites of Estonian organizations, including the Estonian Parliament, banks, ministries, newspapers and broadcasting organisations, etc. Russia was accused of having launched the attacks as retaliation for the relocation of the Soviet statue of the Bronze Soldier in Tallinn.

V. Definition and Scope of the Solidarity Clause

The Solidarity Clause (Art. 222 TFEU) became legally binding on 1 December 2009 with the entry into force of the Treaty of Lisbon. As outlined under II.b.ii., the Solidarity Clause was already included in the provisions of the Constitutional Treaty, albeit divided into two separate Articles. The Treaty of Lisbon merges Articles I-43 and III-329 TCE into one single Article. Although the title 'Solidarity Clause' would suggest a general and horizontal provision on solidarity, Article 222 TFEU remains restricted to specific situations. The Solidarity Clause touches upon different policy areas of the European Union, namely the 'Area of Freedom, Security and Justice' (AFSJ) (Title V of the Third Part of the TFEU), the CFSP (CSDP) and 'Civil Protection' (Article 196, Title XXIII of the Third Part of the TFEU). Probably also the policy areas 'Public Health' (Title XIV), 'Environment' (Title XX) and 'Energy' (Title XXI), which could all be regarded as areas which have (at least) some links to the Clause.

Article 222 TFEU is located in Title VII 'Solidarity Clause' in Part V ('External Action by the Union') of the TFEU. Although the Clause is included under 'External Action', the provision finds itself in a 'grey area' between civilian and military cooperation[121]. This also corresponds to the institutional setting of the anti-terrorism policy of the Union which is split between the AFSJ and the CFSP[122].

V.a. Definition

The Solidarity Clause commits the Union and its Member States to acting jointly in a spirit of solidarity if a Member State is the object of a terrorist

121 See Gregor Schusterschitz, Georg Stillfried, 'Art 222 AEUV', in Heinz Mayer, Karl Stöger, ‚Kommentar zu EUV und AEUV unter Berücksichtigung der österreichischen Judikatur und Literatur' (Wien: Manz Verlag [2012]), (5).

122 See ibid.

attack or the victim of a natural or man-made disaster. Therefore, there are two addressees with regard to the solidarity obligation: The Union and the Member States. However, the Clause distinguishes between both addressees when it comes to the duties and the implementation of the provision[123].

For the purposes of the Solidarity Clause, the Union can mobilise all the instruments at its disposal, including the military resources made available by the Member States, to prevent a terrorist threat in the territory of the Member States and to protect the democratic institutions and the civilian population from any terrorist attack. The Member State concerned, at the request of its political authorities, is to be assisted by the other Member States. To that end, the Member States coordinate between themselves in the Council. As far as the assistance of the EU is concerned, arrangements for the implementation of the Solidarity Clause are to be defined by a decision adopted by the Council acting on a joint proposal by the Commission and the High Representative of the Union for Foreign Affairs and Security Policy. The 'Political and Security Committee' (PSC)[124] and the 'Standing Committee on Internal Security'

---

123 See Christian Calliess, ‚Art. 222 AEUV – Solidaritätsklausel', in Christian Calliess, Matthias Ruffert (Hrsg.), ‚EUV/AEUV – Das Verfassungsrecht der Europäischen Union mit Europäischer Grundrechtecharta' (München: Beck [2011]), 2188.

124 The legal basis of the PSC can be found in Article 38 TEU:
"Without prejudice to Article 240 of the Treaty on the Functioning of the European Union, a Political and Security Committee shall monitor the international situation in the areas covered by the common foreign and security policy and contribute to the definition of policies by delivering opinions to the Council at the request of the Council or of the High Representative of the Union for Foreign Affairs and Security Policy or on its own initiative. It shall also monitor the implementation of agreed policies, without prejudice to the powers of the High Representative.
Within the scope of this Chapter, the Political and Security Committee shall exercise, under the responsibility of the Council and of the High Representative, the political control and strategic direction of the crisis management operations referred to in Article 43.
The Council may authorise the Committee, for the purpose and for the duration of a crisis management operation, as determined by the Council, to take the relevant decisions concerning the political control and strategic direction of the operation."

(COSI)[125] should assist the Council to this end. If necessary, the two committees can submit joint opinions. Furthermore, Article 222 TFEU provides for the European Council to regularly assess the threats facing the Union in order to enable the Union and its Member States to take effective action[126].

V.a.i. Article 222(1) TFEU

"1. The Union and its Member States shall act jointly in a spirit of solidarity if a Member State is the object of a terrorist attack or the victim of a natural or man-made disaster. The Union shall mobilise all the instruments at its disposal, including the military resources made available by the Member States, to:

(a) — prevent the terrorist threat in the territory of the Member States;

— protect democratic institutions and the civilian population from any terrorist attack;

— assist a Member State in its territory, at the request of its political authorities, in the event of a terrorist attack;

(b) assist a Member State in its territory, at the request of its political authorities, in the event of a natural or man-made disaster."[127]

---

125 According to Article 71 TFEU "[a] standing committee shall be set up within the Council in order to ensure that operational cooperation on internal security is promoted and strengthened within the Union. Without prejudice to Article 240, it shall facilitate coordination of the action of Member States' competent authorities. Representatives of the Union bodies, offices and agencies concerned may be involved in the proceedings of this committee. The European Parliament and national Parliaments shall be kept informed of the proceedings."

126 See also the comments of the Austrian delegation (in COSI) on the 'Implementation of Article 222 TFEU (Solidarity Clause)' (10956/12), available at http://www.statewatch.org/news/2013/feb/eu-solidarity-clause-austria-10956–12.pdf http://www.statewatch.org/news/2013/feb/eu-solidarity-clause-austria-10956–12.pdf. The author of the present paper – in his capacity as member of the Austrian delegation in COSI – was involved in the drafting of document 10956/12, which was finally presented in the COSI meeting on 25 June 2012.

127 Article 222(1) TFEU.

Paragraph (1), first sentence, lists three cases, namely a terrorist attack, a natural disaster and a man-made disaster.

V.a.i.a. Terrorist Attack

As regards terrorist attacks: whereas paragraph (1) mentions "a terrorist attack", sentence 2, item (a), also refers to the prevention of "the terrorist threat" (which is even broader than "terrorist attacks") and protection from "any terrorist attack". Neither in EU law nor in international law can a precise definition of the term 'terrorism' or 'terrorist attack' be found[128]. However, as far as EU law is concerned, certain guidance is given by Framework Decision of the Council of 13 June 2002 on combating terrorism (2002/475/JHA) which states that offences that were committed with the aim of seriously intimidating the population, or unduly compelling a Government or international organisation to perform or abstain from performing any act, or seriously destabilising or destroying the fundamental political, constitutional, economic or social structures of a country or an international organisation, are to be deemed to be terrorist offences. According to Framework Decision 2002/475/JHA terrorist offences are:

"(a) attacks upon a person's life which may cause death;
(b) attacks upon the physical integrity of a person;
(c) kidnapping or hostage taking;
(d) causing extensive destruction to a Government or public facility, a transport system, an infrastructure facility, including an information system, a fixed platform located on the continental shelf, a public place or private property likely to endanger human life or result in major economic loss;
(e) seizure of aircraft, ships or other means of public or goods transport;
(f) manufacture, possession, acquisition, transport, supply or use of weapons, explosives or of nuclear, biological or chemical weapons, as well as research into, and development of, biological and chemical weapons;

---

128 On the definition of 'terrorism' cf. Gerhard Hafner, 'The Definition of the Crime of Terrorism', in Giuseppe Nesi, 'International Cooperation in Counter-terrorism' (Ashgate Publishing [2006]), 33–43.

(g) release of dangerous substances, or causing fires, floods or explosions the effect of which is to endanger human life;
(h) interfering with or disrupting the supply of water, power or any other fundamental natural resource the effect of which is to endanger human life;
(i) threatening to commit any of the acts listed in (a) to (h)."[129]

Working Group VIII mentioned in its final report that "[s]uch a clause would not be a clause on collective defence entailing an obligation to provide military assistance, but would apply to threats from non-State entities"[130]. Does this automatically mean that the Solidarity Clause only applies to terrorist attacks from non-State entities as Calliess[131] argues, for example, or could it also be applicable to terrorist attacks which may trigger individual and collective self-defence in accordance with Article 51 UN Charter?

With regard to the latter we may recall that UN Security Council Resolution 1368(2001) condemned the terrorist attacks of 9/11 and affirmed that it would regard "such acts, like any act of international terrorism, as a threat to international peace and security"[132]. As Article 42(7) TEU explicitly refers to Article 51 UN Charter, we may come up with situations where both provisions could apply. Calliess finds a solution to this possible overlap by referring to the final report of the Working Group VIII where – as we have seen – the notion of terrorist attacks is confined to a situation where non-State entities carry out the terrorist attack. According to Calliess, Article 222 TFEU is – compared to Article 42(7) TEU – the *lex specialis*[133]. Calliess deduces this interpretation from the fact that the Convention[134] had already affirmed that the Solidarity Clause would only apply to attacks from non-State entities. Therefore, terrorist attacks committed by a third State would only trigger the Mutual

129 Official Journal of the European Union L 164/4, 'Framework Decision of the Council of 13 June 2002 on combating terrorism (2002/475/JHA)', 22.6.2002.
130 'Final Report WG VIII', 20.
131 See Calliess, 2189.
132 S/RES/1368 (2001) of 12 September 2001.
133 See Calliess, 2189.
134 See CONV 685/03, 22.

Assistance Clause[135]. This interpretation is not completely convincing. If Article 222 TFEU is – in the case of a terrorist attack – a *lex specialis*, following the principle *lex specialis derogat legi generali* it should be applied instead of Article 42(7) TEU and irrespective of who committed the terrorist attack (which, of course, is not always clear in the moment a terrorist attack has occurred). The reference to the explanations of the Praesidium of the Convention (CONV 685/03) and the Final Report of the Working Group VIII raises the question of the legal relevance of these 'interpretations'. According to the rules of interpretation enshrined in Article 31 of the Vienna Convention on the Law of the Treaties (VCLT)[136], the textual interpretation is crucial. The preparatory work of the WG VIII and the note of the Praesidium of the Convention come under the meaning of *travaux préparatoires* and would, therefore, be a supplementary means of interpretation in accordance with Article 32 VCLT. Recourse to a supplementary means of interpretation may be had "in order to confirm the meaning resulting from the application of article 31, or to determine the meaning when the interpretation according to article 31: (*a*) leaves the meaning ambiguous or obscure; or (*b*) leads to a result which is manifestly absurd or unreasonable"[137]. Referring to the text of Article 222 TFEU, we cannot derive any limitation of the notion of 'terrorist attacks' to terrorist attacks committed by non-State entities. According to Article 31(4) VCLT, a "special meaning shall be given to a term if it is established that the parties so intended". The commentaries to the VCLT outline that Article 31(4) VCLT "provides for the somewhat exceptional case where, notwithstanding the apparent meaning of a term in its context, it is established that the parties intended it to have a special meaning"[138]. In our case, the term 'terrorist attacks' should be complemented with a reference to 'non-State entities'. This special meaning could be then sought from the interpretation included in the note of the Praesidium of the Convention and

---

135 See Calliess, 2189.

136 Vienna Convention on the Law of Treaties of 23 May 1969, UNTS I-18232.

137 Article 32 VCLT.

138 Draft Articles on the Law of Treaties with commentaries, available at http://untreaty.un.org/ilc/texts/instruments/english/commentaries/1_1_1966.pdf, 37.

the final report of WG VIII. It could, however, be more coherent to apply Article 32 VCLT, which would allow reference to "the preparatory work of the treaty and the circumstances of its conclusion, in order to confirm the meaning resulting from the application of article 31"[139]. As the preparatory work done by WG VIII seems to clearly indicate the intention to restrict the notion of terrorist attacks to attacks perpetrated by non-State entities, we may come to the same conclusion as Calliess. Some doubts, certainly, still remain; in particular, if we consider the rather scarce and vague explications of the WG VIII on the "threats from non-State entities"[140]. As both Clauses have not yet been invoked we can only speculate which Clause, if any, would be invoked after a terrorist attack has occurred. It would, however, appear more reasonable if first and foremost the Solidarity Case – as *lex specialis* with regard to terrorist attacks – is invoked in the case of an apparent terrorist offence, independent of the nature and origin of the perpetrator(s).

V.a.i.b. Terrorist Threat

Article 222(1), first sentence, apparently restricts the scope of the Clause to three cases: terrorist attack, natural and man-made disaster. Paragraph (1)(a), first indent, extends the scope to the "terrorist threat": The Union is obliged "to prevent the terrorist threat in the territory of the Member States". This means that the Union must perform preventive tasks which may only be performed in order to avoid terrorist attacks "in the territory of the Member States". Threatening to commit any crimes that are listed in Article 1(a) to (h) of Framework Decision (2002/475/JHA) is to be considered a terrorist offence. Therefore, we can also rely on the catalogue of offences listed in the Framework Decision on combating terrorism for the definition of the terrorist threat. The question arises whether, and if so, how the Union could act preventively at all and in particular whether this would be possible without interfering with the sovereignty of Member States?

---

139 Article 32 VCLT.

140 Final Report, 20.

In response to the terrorist attacks of 9/11, Madrid and London, the European Union has adopted several strategic documents that aim to foster cooperation between Member States as well as to improve the Union's own capacities to prevent and combat international terrorism. The European Security Strategy of 2003 focuses on preventive action by the Union: "The risks of proliferation grow over time; left alone, terrorist networks will become ever more dangerous. State failure and organised crime spread if they are neglected – as we have seen in West Africa. This implies that we should be ready to act before a crisis occurs. Conflict prevention and threat prevention cannot start too early."[141]

The European Union Counter-Terrorism Strategy of 2005 is, amongst other things, based on a 'prevention pillar' which should contribute to combat the radicalisation and recruitment of terrorists by identifying the methods, propaganda and the instruments used by terrorists[142]. Despite these and other documents (such as the European Union Strategy for Combating Radicalisation and Recruitment to Terrorism)[143], the means available to the Union remain limited due to the lack of competence of the Union as regards national security (Article 4 (2) TFEU)[144] and internal security (Article 72 TFEU). By and large, we can mention three main actors in the Union that deal with Counter-Terrorism: The Counter-Terrorism-Coordinator (CTC) of the Council, Europol and the EU Intelligence Analysis Centre (IntCen).

The CTC was established by the European Council after the bombings that took place on 11 March 2004 in Madrid. In the declaration of 25 March 2004 the heads of State and government of the EU defined the tasks of the CTC:

> "The Co-ordinator, who will work within the Council Secretariat, will co-ordinate the work of the Council in combating terrorism and, with due regard to the responsibilities of the Commission, maintain an overview of all the instruments at the Union's

141 'A secure Europe in a better World - European Security Strategy', 8.

142 'The European Union Counter-Terrorism Strategy' (14469/4/05 Rev 4).

143 'The European Union Strategy for Combating Radicalisation and Recruitment to Terrorism' (14781/1/05).

144 Article 4(2) TFEU: "In particular, national security remains the sole responsibility of each Member State."

disposal with a view to regular reporting to the Council and effective follow-up of Council decisions"[145].

According to the declaration of 25 March 2004, the role and the tasks of the CTC are, by and large, of a merely administrative and coordinative nature. The CTC should also be perceived as the 'face' of the Union when it comes to counter-terrorism matters both internally and with regard to external relations. This 'public relations' task contributes to the representative role of the CTC which has gained in importance over the years. Moreover, the CTC may table reports on issues of particular concern (such as the challenges posed by 'foreign fighters')[146] to the Council and make recommendations. However, the tasks of the CTC exclude any specific (operational) role in the implementation of the Solidarity Clause and in particular in the Union's obligation to prevent terrorist threats.

The European Police Office is more likely to be a suitable addressee of the preventive tasks provided for in Article 222 TFEU. According to Article 88(1) TFEU, "Europol's mission shall be to support and strengthen action by the Member States' police authorities and other law enforcement services and their mutual cooperation in preventing and combating serious crime affecting two or more Member States, terrorism and forms of crime which affect a common interest covered by a Union policy." Certainly, the core tasks of Europol lay in the area of organised crime (OC) but some instruments have also been established in the counter-terrorism field, such as the 'Counter Terrorism Task Force'. The Counter Terrorism Task Force was established after 9/11 and comprised experts and liaison officers from Member States. It was discontinued after the immediate terrorist threat was considered to have subsided, but reactivated after the Madrid train attacks in March 2004[147]. It is also worth mentioning the 'Check

145 European Council, 'Declaration on combating terrorism', 25 March 2004, 13.

146 Concerning the threats posed by fighters from Europe crossing the border into Syria to fight, the Council held two debates in June and December 2013 which were based on a report drafted by the CTC (see the press release of the 3279th Council meeting of the Justice and Home Affairs ministers held in Brussels, 5 and 6 December 2013, Council document 17342/13).

147 See Andy Brown, 'Ten Years of Europol 1999 – 2009', available at https://www.europol.europa.eu/sites/default/files/publications/anniversary-publication.pdf, 25.

the Web' application aimed at monitoring Islamist terrorism on the web, and ATLAS, a specialised law enforcement portal to enhance collaboration between anti-terrorism teams[148]. Finally, we should also mention the 'EU Terrorism Situation and Trend Report' (TE-SAT) first published in 2007. Unlike the 'Serious and Organised Crime Threat Assessment' (SOCTA)[149] which aims to provide decision-makers with a document on which to base their medium and longer term strategic planning, TE-SAT is a retrospective assessment and informs the public on the state of terrorism in the EU. Despite the rather cautious approach shown by Europol in counter-terrorism matters so far (an approach which was surely due to the traditional reluctance in particular of the large Member States to confer more influence on and share information with Union bodies in matters that relate to national security), more activities in the analysis of the potential threat could be a direct consequence of the implementation of the Solidarity Clause. Operational preventive tasks performed by Europol 'on the ground' without being embedded in a Joint Investigation Team[150] remain, however, excluded by Europol's legal basis[151].

The 'EU Intelligence Analysis Centre' (IntCen) is an intelligence body and part of the European External Action Service (EEAS). It was established 2011 but already existed before as 'Situation Centre' (SITCEN) within the General

148 See ibid., 47.

149 The EU Serious and Organized Crime Threat Assessment (SOCTA) was first adopted in 2013 and is the product of systematic analysis of law enforcement information on criminal activities and groups affecting the EU. The SOCTA is designed to assist strategic decision-makers in the prioritization of organized crime threats. The public version of SOCTA is available at https://www.europol.europa.eu/sites/default/files/publications/socta2013.pdf

150 A Joint Investigation Team (JIT) is an investigation team set up for a fixed period, based on an agreement between two or more EU Member States and/or competent authorities, for a specific purpose. Non-EU Member States may participate in a JIT with the agreement of all other parties. JITs investigate specific cases (see https://www.europol.europa.eu/content/page/joint-investigation-teams-989).

151 Article 88(3) TFEU: "Any operational action by Europol must be carried out in liaison and in agreement with the authorities of the Member State or States whose territory is concerned. The application of coercive measures shall be the exclusive responsibility of the competent national authorities."

Secretariat of the Council. IntCen's mission is to provide intelligence analyses, early warnings and situational awareness to the High Representative and to the European External Action Service, to the various EU decision-making bodies in the fields of the CFSP, CSDP and counter-terrorism, as well as to the EU Member States. IntCen accomplishes this mission by monitoring and assessing international events 24 hours a day, 7 days a week, focusing particularly on sensitive geographical areas, terrorism, the proliferation of weapons of mass destruction and other global threats[152]. IntCen is still a very small entity and employs around seventy people. There are two main divisions within IntCen: the Analysis Division, which is split into sections based on geographical and thematic topics, and the General and External Relations Division dealing with legal, administrative and IT issues and undertaking open source analysis[153]. Like Europol, IntCen can also provide threat assessments which could be used for the purposes of Article 222(1)(a), first indent, although also in this case no operational intervention by IntCen itself – quite apart from the fact that it does not have enough appropriate staff for this task – would be possible without interfering in sovereign rights of the affected Member State.

V.a.i.c. Natural Disaster

Protection from "any terrorist attack" may indicate that in the case of a terrorist attack no threshold requirement is to be considered when applying the Clause. Indeed, this approach would be also in line with the Framework Decision of the Council of 13 June 2002 on combating terrorism (2002/475/JHA), which does not restrict the notion of 'terrorist attack' to attacks of a certain gravity. Does this *e contrario* mean that for natural or man-made disasters it could be possible to limit the obligation of the Union and the Member States arising out of the Clause to major disasters?

---

152 See the statement of the first director of Ilkka Salmi, Director of the Situation Centre of the European Union, available at http://www.europarl.europa.eu/meetdocs/2009_2014/documents/sede/dv/sede041011cvsalmi_/sede041011cvsalmi_en.pdf.

153 See Chris Jones, 'Secrecy reigns at the EU's Intelligence Analysis Centre' (Statewatch Analysis), available at http://www.statewatch.org/analyses/no-223-eu-intcen.pdf.

The Council Regulation of 11 November 2002 establishing the European Union Solidarity Fund[154] confines assistance to Member States under that instrument "to major disasters with serious repercussions on living conditions, the natural environment or the Economy"[155] (fifth indent). According to this Council Regulation "a 'major disaster' within the meaning of this Regulation should mean any disaster, in at least one of the States concerned, resulting in important damage expressed in financial terms or as a percentage of the gross national income (GNI)"[156]. However, assistance is not to be limited to disasters causing major financial damage. "In order to permit interventions in the case of disasters that, while important, do not reach the minimum scale required, assistance may also be granted under exceptional circumstances in case an eligible neighbouring country is affected by the same disaster, or whenever the major part of the population of a specific region is affected by a disaster with serious and lasting repercussions on living conditions"[157] (sixth indent).

In this respect it may be advisable to follow the line drawn by the work done by the *Institut de Droit International* with its resolution on 'Humanitarian Assistance'[158] or to look at the 'Text of the draft articles on the Protection of persons in the event of disasters', as yet provisionally adopted by the International

154 Official Journal of the European Union L 311/3, 14.11.2002.

155 Ibid., 1

156 Ibid.

157 Ibid.

158 See the Resolution on 'Humanitarian Assistance' of the Institut de Droit International (Institute of International Law) in its Bruges 2003 session (available at http://www.idi-iil.org/idiE/resolutionsE/2003_bru_03_en.PDF):

"2. "Disaster" means calamitous events which endanger life, health, physical integrity, or the right not to be subjected to cruel, inhuman or degrading treatment, or other fundamental human rights, or the essential needs of the population, whether

- of natural origin (such as earthquakes, volcanic eruptions, windstorms, torrential rains, floods, landslides, droughts, fires, famine, epidemics), or
- man-made disasters of technological origin (such as chemical disasters or nuclear explosions), or
- caused by armed conflicts or violence (such as international or internal armed conflicts, internal disturbances or violence, terrorist activities)."

Law Commission[159], as well as at the Memorandum of the UN Secretariat, 'Protection of Persons in the Event of Disasters'[160]. A higher threshold requirement for natural and man-made disasters cannot be derived from the preparatory work of Working Group VIII, which does not even mention the possibility of natural or man-made disasters as possible causes triggering the activation of the Solidarity Clause[161].

Article 222 TFEU applies "in the event of a natural or man-made disaster" without any reference to preventive action, as is the case with terrorist threat and terrorist attacks. Therefore, it could be assumed that in the case of natural or man-made disasters preventive measures are outside the scope of the Solidarity Clause.

Article 222 TFEU (1)(a), third indent, as well as (b), require a request for assistance from the "political authorities" of the affected Member States. The first and second indents of item (a), however, do not contain such a requirement. Is the Union allowed to act without prior request from the affected Member State? If we consider the general rules of interpretation of the Vienna Convention on the Law of the Treaties, the "ordinary meaning to be given to the terms"[162] does not contradict this assumption nor does the context of the terms „in the light of its object and purpose“[163]. If we look at the third indent,

---

159 See the 'Report of the International Law Commission Sixty-third session (26 April–3 June and 4 July–12 August 2011)', In: General Assembly Official Records Sixty-sixth session Supplement No. 10 (A/66/10): Article 3 'Definition of disaster'
"Disaster" means a calamitous event or series of events resulting in widespread loss of life, great human suffering and distress, or large-scale material or environmental damage, thereby seriously disrupting the functioning of society.

160 International Law Commission Sixtieth session Geneva, 5 May-6 June and 7 July-8 August 2008, 'Protection of persons in the event of disasters', Doc A/CN, 4/590.

161 Schusterschitz, Stillfried take it for granted – without further explanation – that natural or man-made disasters must meet a higher threshold requirement: "Anders als bei der Qualifikation eines Anschlages als terroristisch, muss aber das Ereignis ein höheres Schadensausmaß für die Umwelt bzw das Leben und die Gesundheit von Menschen oder Tieren überschreiten, um als Katastrophe qualifiziert zu werden" (Schusterschitz, Stillfried, (15)).

162 Article 31(1) VCLT.

163 Ibid.

we could also affirm that *e contrario* not mentioning a request from political authorities means that the Union may or even should take action without any prior request from (or even tacit consent of?) the affected Member State. Certainly, this raises a number of difficult questions with regard to the sovereignty of States, in particular if we consider that the Union may also use military means made available by the Member States. Unfortunately, the Commission and the High Representative missed a good opportunity to clarify this issue in their 'Joint Proposal' on the implementation of the Solidarity Clause by the Union (see under VI.). It should indeed be clarified whether all cases falling under Article 222(1) and (2) require a request from the political authorities of the affected Member States for assistance, or, whether the range of measures available to the EU in cases where the assent of the Member State is not mentioned (paragraph (1) item (a) first and second indent) is restricted to such measures not infringing upon the sovereignty of the Member State concerned. Moreover, the question should be considered whether the Union could also act in the absence of a request by a Member State (e. g. based on an EU threat assessment)[164].

With regard to the 'preventive scenario' of paragraph (1), item (a), first and second indent, the question arises as to whether the preventive tasks of the Union enshrined in paragraph (1), item (a), first and second indent, require a terrorist attack to have already occurred. In this regard the different wording of the first and second indent should be considered. Whereas the first indent says that the Union is to prevent "the terrorist threat" the second indent provides for an obligation of the Union to protect itself from "any terrorist attack". The first relates to a specific threat, whereas the second indent points at the general possibility of a terrorist attack[165].

V.a.i.d. Man-Made Disaster

The extension of the scope of the Clause to include man-made disasters can be interpreted with the intention of the authors of Article 222 to focus on the

164 See 10956/12, 3.

165 See ibid., 6.

effects of a disaster rather than on its origin. Disasters such as forest fires, nuclear incidents or floods can be either of natural origin or they can be caused by, for example, human error. According to Calliess, the Solidarity Clause should only be triggered in the case of a major disaster, also with regard to man-made disasters[166]. The text of the Clause, however, does not limit assistance to cases of 'major' disasters. In order to have a clearer picture of what the Treaty could mean by the notion of 'man-made disaster', let us turn to the so-called 'Barnier Report'[167] of 2006. The Barnier Report lists several and quite diverse examples of (natural and) man-made disasters, such as the Darfur crisis, the maritime accident of the oil tanker 'Prestige', the terrorist attacks of 9/11 or the humanitarian crisis in Somalia[168].

The 'Council Decision of 8 November 2007 establishing a Community civil protection mechanism (CPM) (recast)' states in its Article 1(2):

> "2. The protection to be ensured by the Mechanism shall cover primarily people but also the environment and property, including cultural heritage, in the event of natural and man-made disasters, acts of terrorism and, technological, radiological or environmental accidents, including accidental marine pollution, occurring inside or outside the Community, taking also into account the special needs of the isolated, outermost and other regions or islands of the Community."[169]

Neither the Barnier Report nor the CPM clearly outline the term 'man-made disaster'. It is, for example, questionable whether war could also be subsumed under this term. The Working Group VIII did not tackle this question and its final report only refers to "natural or humanitarian disasters within the

166 See Calliess, 2190.

167 On 9 May 2006, former European Commissioner Michel Barnier produced the report 'For a European civil protection force: Europe Aid' on European civil protection commissioned by the President of the Commission. The report centres around the creation of a European civil protection force ('Europe Aid'). The Force would also be able to acquire additional resources at EU level. The focal point of the Force would be an Operations Centre (see http://ec.europa.eu/echo/civil_protection/civil/prote/perspectives_en.htm).

168 Barnier Report, 10.

169 Council Decision 8 November 2007 establishing a Community civil protection mechanism (Official Journal of the European Union L 314/9, 1.12.2007).

Union"[170]. Certainly, there can be no doubt about the human origin of war and about the fact that any war is potentially a major disaster. The distinction made by WG VIII between natural and humanitarian disasters does not add much to clarify the scope of the term 'man-made disaster'. It could be argued that by using the term 'humanitarian' instead of man-made the WG VIII wanted to stress the consequences and not the origin of the disaster. However, a natural disaster could also turn out to be a humanitarian crisis. War or the consequences of war will regularly cause a humanitarian crisis, as was the case in the recent past in Somalia or in Darfur.

All in all, the exclusion of (civil) war from the scope of the Clause may be consistent with the original aim of the Clause, although this was first and foremost oriented at preventing and responding "to terrorist threats within the Union by mobilising all the necessary military and civil instruments"[171]. The question is rather what would happen if, for example, in a situation of violent unrest or even civil war in a Member State the official government triggers the Solidarity Clause. Although this may still be a merely hypothetical situation, it would be difficult to affirm that on the one hand the Union "shall promote [...] solidarity among Member States" (Art. 3(3) TEU) but on the other hand it can (legally) accept complete non-assistance in a situation of major crisis within the territory of a Member State. The inclusion of war under the scope of the Solidarity Clause could, of course, have relevant political and legal consequences, in particular with regard to the status of neutral and non-aligned Member States.

V.a.i.e. Forms of Assistance

Article 222(1), second sentence, enumerates three forms of assistance that the Union (and not the Member States) may render to the affected Member States. The Union shall

"— prevent the terrorist threat in the territory of the Member States;
— protect democratic institutions and the civilian population from any terrorist attack;

170 Final Report Working Group VIII, 21.
171 Ibid., 2.

— assist a Member State in its territory, at the request of its political authorities, in the event of a terrorist attack".

This part of the Solidarity Clause is somehow the most puzzling. As outlined under V.a.i.b., the prevention of the terrorist threat is restricted to the territory of the Member States. The first indent is the only provision where we find the term 'terrorist threat', which – as already outlined – extends the scope of the Clause and introduces an obligation for the Union to act preventively. Although the second indent goes back to the notion of 'terrorist attack', it appears to keep a preventive remit in referring to an obligation to "protect [...] from any terrorist attack". Protecting from any terrorist attack does not necessarily mean that the relevant incident has already occurred. On the contrary, in this case the obligation on the Union also seems rather oriented towards the prevention of harm to "democratic institutions and the civilian population" in case a terrorist attack occurs. This act of protection would only apparently be physical protection, which, of course, the Union would hardly be able to guarantee, but may, for example, result in the adoption of appropriate legal provisions with regard to security in the field of aviation in order to protect the civilian population travelling by plane (irrespective of the territory) from a hypothetical (therefore, the word 'any') terrorist attack. Of course, the problem with this interpretation is that the first sentence of Article 222(1) establishes the general rule that the Solidarity Clause only applies "if a Member State is the object of a terrorist attack or the victim of a natural or man-made disaster". Therefore, a certain kind of probability that a Member State could be the 'object' of a terrorist attack in the near future will be necessary to apply the Solidarity Clause, also with regard to the first and second indent. However, the wording of the second indent remains ambiguous as no reference to a Member State or to its territory can be found.

V.a.i.f. Democratic Institutions and the Civilian Population

According to Article 222(1)(a), second indent, the Union is obliged to protect the democratic institutions and the civilian population from any terrorist attack. Although the provision lacks any limitation to the territory of the Member States, there can be no doubt that the protection by the Union applies to the democratic institutions and the civilian population of Member States[172]. It is, however, a more uncertain issue which institutions are to be considered as 'democratic institutions' under the Solidarity Clause. Democratic institutions are typically institutions that can be found in democratic countries such as parliaments (national and regional), seats of governments, courts, etc. However, even in non-democratic countries we can find the same institutions. As all EU Member States are and have to be democratic States in order to be EU Member States, the reference to 'democratic' institutions is probably the attempt to restrict the protection to those institutions that are the pillars of a democratic system of government. Under this restrictive view we could subsume parliaments, the seat of the government and the seat of the different ministries (as the ministers and the prime minister/head of State together form a government) and also the judiciary institutions (prosecution and courts). Therefore, institutions such as the court of auditors or embassies in third countries would not be included under this provision as they are not strictly necessary for the maintenance of a democratic system. Certainly, if the Clause is triggered, it would not make much sense to distinguish between institutions that are vital to democratic life and others which are also important but not indispensable. Thus, a broader approach encompassing all State institutions serving the civilian population (including local and regional institutions, courts, courts of audit, national banks, embassies, offices of ombudspersons, etc.) may be more reasonable.

The term civilian population is *per se* less ambiguous. Military forces of a Member State would, therefore, not enjoy the protection of the Union[173]. Could, for example, the crew of a (national) broadcasting company or the staff of a (national) enterprise enjoy the protection of the Clause? A definition

172 See Schusterschitz, Stillfried, (10).

173 According to Webster dictionary a civilian is a person not on active duty in the armed services or not on a police or fire fighting force, see at http://www.merriam-webster.com/dictionary/civilian.

of 'civilian' can be found in humanitarian law, specifically in Article 50 of the 'Protocol Additional to the Geneva Conventions of 12 August 1949, and relating to the Protection of Victims of International Armed Conflicts (Protocol I) of 8 June 1977'[174]:

> "A civilian is any person who does not belong to one of the categories of persons referred to in Article 4 A 1), 2), 3) and 6) of the Third Convention and in Article 43 of this Protocol. In case of doubt whether a person is a civilian, that person shall be considered to be a civilian. A civilian is any person who does not belong to one of the categories of persons referred to in Article 4 A 1), 2), 3) and 6) of the Third Convention and in Article 43 of this Protocol. In case of doubt whether a person is a civilian, that person shall be considered to be a civilian."

Therefore, a civilian is not a combatant in accordance with the Geneva Conventions, which confirms the dictionary interpretation. Hence, the 'civilian population' encompasses a very wide range of nationals of a Member State. As regards the duties of the Union it is, therefore, crucial to establish whether not mentioning a territorial limitation means that the protection extends to civilians and democratic institutions in third countries. Once again, with regard to this question the European Commission and the High Representative missed a good opportunity to provide a sound interpretation of the scope of the Clause when drafting the Joint Proposal on the implementation of the Clause (see chapter VI.).

V.a.i.g. Political Authorities

The request for assistance must be made by the "political authorities" of the affected Member State. The term 'political' excludes merely administrative authorities such as emergency centres within the framework of the civil protection system of a Member State. The reference to the political authorities of a Member State could also be interpreted as indication that the implementing

174 'Protocol Additional to the Geneva Conventions of 12 August 1949, and relating to the protection of victims of international armed conflicts (Protocol I)', UNTS I-17512.

provisions of the Clause to be adopted by the Council in accordance with Article 222(3) TFEU must also include a political statement of the EU institutions. We will see in chapter VI. that the Joint Proposal inadequately takes into account the political dimension of the Clause.

According to Vedder, the notion of 'political authorities' goes beyond what international law considers to be persons representing a State[175]. Political authorities will most likely be the head of State, members of the government and parliament. Possibly also regional political authorities could fall under this provision. However, in order to guarantee that the 'right' authorities call for assistance under the Solidarity Clause, it would be advisable to adopt a list of competent authorities, for instance as an annex to the decision adopted by the Council under Article 222(3) TFEU.

V.a.ii. Article 222(2) TFEU

"2. Should a Member State be the object of a terrorist attack or the victim of a natural or man-made disaster, the other Member States shall assist it at the request of its political authorities. To that end, the Member States shall coordinate between themselves in the Council."[176]

Before we analyse this provision we should recall that in the TCE the Clause was split into two separate Articles. In Article III-329(1) TCE the arrangements for implementing the Solidarity Clause contained in Article I-43 referred to the duty of the other Member States to assist the affected Member State at the request of its political authorities. The text of Article III-329(1) TCE is the same as Article 222(2) TFEU. The fact that the Lisbon Treaty merges the two provisions of the TCE creates confusion as regards the scope of paragraph (2). As already mentioned, paragraph (2) lacks any territorial limitation. This could lead us to the conclusion that whenever a Member State is the object of a terrorist attack or the victim of a natural or man-made disaster and a request is made by the political authorities, the Member States, regardless

175 See Vedder, 842.

176 Art. 222(2) TFEU.

of the territory concerned, are obliged to assist. The territorial limitations would then only apply to the Union, which would have, if any, very limited means to assist a Member State outside of its territory. Nevertheless, the reference to the invocation of the Clause by the political authorities (similarly to paragraph (1)(a), third indent, and paragraph (1)(b)) could also be interpreted as limitation to assistance on the territory of the affected Member State[177].

However, this interpretation is not completely convincing and, taking into account the geographical scope adopted by the Joint Proposal, it appears as if at least the Commission and the High Representative are not following this narrow interpretation. If we consider the original TCE structure of the Solidarity Clause, we could also come to the conclusion that Article 222(1), first sentence, sets an overall obligation for the Member States which is not limited from a geographical point of view. Recalling the structure of the TCE, Article 222(2) is only part of the implementing provision (Article III-329 TCE) and hence merely specifies that the Member States are only to provide assistance upon request of the political authorities of the affected Member States and are to coordinate between themselves in the Council. Paragraph (2) is, therefore, a merely procedural provision. The substantive elements are already included in Article 222(1). The fact that the assistance of the Union is partly limited to the territory of the Member States does not mean that the limitation must also apply to the Member States.

It is also questionable whether the cases triggering solidarity in paragraph (2) are the same as in paragraph (1), strictly speaking the 'preventive' cases in paragraph (1)(a). As the different sub-cases mentioned in paragraph (1) exclusively relate to the Union's obligation, and the obligation upon the Member States clearly refers to a situation where the terrorist attack or the natural or man-made disaster has already occurred, there is no reason to extend the Member States' obligation to assist in preventive cases, too. Of course, from a practical and political point of view there could be exceptional situations where the core values on which the European Union is founded are challenged by the action of a Member State. For such a case the Union has only two

177 See Schusterschitz, Stillfried, (12).

instruments: infringement proceedings and the Article 7 TEU 'sanctions'. Both instruments may turn out to be inefficient, in particular when timely action is imperative to avoid greater harm to the civilians of the affected Member State and maybe also other Member States. The application of the Solidarity Clause could be considered as instrument of last resort although, of course, preventive intervention by the Union and/or the Member States even in exceptional circumstances would challenge the consistency of this kind of action with international law.

Another remarkable difference between paragraph (1) and (2) is not mentioning the military means of assistance. In this case we also have to consider the original structure of the Solidarity Clause in the TCE. The assistance of the Member States is an assistance *tout court* and is only limited by the declaration no. 37 annexed to the Final Act of the Lisbon Intergovernmental Conference which adopted the Treaty of Lisbon. After the rejection of the Constitutional Treaty by France and the Netherlands in 2005, a period of reflection and treaty revision followed. However, the reflection phase did not affect the wording of the Solidarity Clause, with one exception: a joint declaration, originally formulated in the Solidarity Declaration, was added to the Clause to give Member States more latitude in deciding how they would provide assistance[178]. Declaration no. 37 reads as follows:

> "Without prejudice to the measures adopted by the Union to comply with its solidarity obligation towards a Member State which is the object of a terrorist attack or the victim of natural or man- made disaster, none of the provisions of Article 222 is intended to affect the right of another Member State to choose the most appropriate means to comply with its own solidarity obligation towards that Member State."

Member States are free to use the most appropriate means irrespective of the measures adopted by the Union and probably also irrespective of specific requests from the Member States. Nevertheless, this freedom of choice is subject to the limitations set out by the Treaties. It would therefore be difficult

178 See Sara Myrdal, Mark Rhinard, 'The European Union's Solidarity Clause: Empty Letter or Effective Tool? An Analysis of Article 222 of the Treaty on the Functioning of the European Union' (Swedish Institute of International Affairs [2010]), 6.

to consider the solidarity principle fulfilled if, for instance, a Member State affected by a CBRN attack invokes the Clause and a neighbouring Member State which compared to other Member States would have the most appropriate (technical) equipment for this kind of disaster limits its assistance to a political declaration. The principle of solidarity[179] would obviously be jeopardized and the Solidarity Clause itself would lose any practical relevance for possible future emergency situations.

Certainly, there is no limitation to the means to be used by the assisting Member States. Hence, also military means could be used. An obligation to assist with military means cannot be derived from the wording of the Clause[180]. As regards possible limitations of the assistance with military means we should also take into consideration that according to UN General Assembly (GA) Resolution 3314 (XXIX) "[t]he use of armed forces of one State which are within the territory of another State with the agreement of the receiving State, in contravention of the conditions provided for in the agreement or any extension of their presence in such territory beyond the termination of the agreement"[181] qualifies as an act of aggression[182]. Taking into account this provision, it would be necessary to already circumscribe assistance in the request of the affected Member State. It is noteworthy that the International Court of Justice (ICJ) considers parts of the GA Resolution 3314 (XXIX) to reflect customary international law[183]. In the Nicaragua case the ICJ also states "that it is the State which is the victim of an armed attack which must form and declare the view that it has been so attacked. There is no rule in customary

179 See on the principle of solidarity Araceli Mangas Martín, Diego J. Liñán Nogueras, 'Institutiones y Derecho de la Unión Europea' (Madrid: Editorial Tecnos [2012]), 59–60.

180 See Calliess, 2196.

181 A/RES/3314(XXIX) of 14 December 1974, Art. 3(e).

182 See Hafner, ‚Beistands- und Solidaritätsklausel', 402.

183 See ICJ, Reports of Judgments, Advisory opinions and orders, 'Case concerning military and paramilitary activities in and against Nicaragua (Nicaragua v. United States of America)', 103. The ICJ says that the description of armed aggression "contained in Article 3, paragraph (g), of the Definition of Aggression annexed to General Assembly resolution 3314 (XXIX), may be taken to reflect customary international law".

international law permitting another State to exercise the right of collective self-defence on the basis of its own assessment of the situation. Where collective self-defence is invoked, it is to be expected that the State for whose benefit this right is used will have declared itself to be the victim of an armed attack."[184] Customary international law provides for the need to invoke assistance. This is reflected in the Solidarity Clause in the cases where the Clause is triggered upon request of the political authorities. However, it remains questionable whether, once the affected Member State has invoked the Clause without limiting assistance to certain means, it could limit assistance *ex post* or argue that the assisting Member States have infringed its sovereignty. The invocation of the Solidarity Clause could be interpreted as an enabling clause giving the right of intervention. This right would only be limited by the principles of subsidiarity, necessity and proportionality.

With regard to the role of the Council as a mandatory platform for coordination ("[...] shall coordinate between themselves in the Council"), the Clause remains vague concerning the question whether the Member States are to coordinate on a case-by-case basis or whether a horizontal measure, as provided for in Art. 222(3), is needed[185]. Nevertheless, recalling the wording of the first sentence of paragraph (2) ("[s]hould a Member State be the object of a [...]"), it appears as if coordination should rather take place every time the Solidarity Clause has been invoked. Moreover, there is already a coordination mechanism in place, namely the 'Crisis Coordination Arrangements' (CCA). The CCA facilitates cooperation among Member States at a political level when taking strategic decisions in the case of "major emergencies inside or outside the European Union"[186]. The CCA, which have recently been reviewed and renamed as 'Integrated Political Crisis Response arrangements' (IPCR), are "dedicated to facilitating intergovernmental decision making"[187]. The EU Integrated

184 Ibid., 104.

185 Schusterschitz, Stillfried, (12).

186 Council of the European Union (9409/06, Presse 144), 28.

187 Marie Fuchs-Drapier, 'The European Union's Solidarity Clause in the Event of a Terrorist Attack: Towards Solidarity or Maintaining Sovereignty?' (Blackwell Journal of Contingencies and Crisis Management, Volume 19 Number 4 [December 2011]), 189.

Political Crisis Response arrangements reinforce the European Union's ability to take rapid decisions when facing major emergencies requiring a response at EU political level. They were approved on 25 June 2013 by the Council. They also support the Solidarity Clause and involve not only the Council but also the Commission, the EEAS and the General Secretariat of the Council (GSC) who, among other things, manages the IPCR Web Platform[188].

Due to the mandatory involvement of the EU institution of the 'Council' and the fact that also other EU bodies are involved in the coordination arrangements of the Member States, the provision of paragraph (2), second sentence, cannot be interpreted as being merely intergovernmental[189]. On the contrary, it could be even interpreted as highlighting the primacy of European Union mechanisms of coordination (such as the CCA/IPCR) with respect to relevant international law provisions (*lex specialis* rule)[190].

V.a.iii. Article 222(3) TFEU

"3. The arrangements for the implementation by the Union of the solidarity clause shall be defined by a decision adopted by the Council acting on a joint proposal by the Commission and the High Representative of the Union for Foreign Affairs and Security Policy. The Council shall act in accordance with Article 31(1) of the Treaty on European Union where this decision has defence implications. The European Parliament shall be informed.

For the purposes of this paragraph and without prejudice to Article 240, the Council shall be assisted by the Political and Security Committee with the support of the structures developed in the context of the common security and

188 See Council of the European Union, IPCR Secretariat, 'IPCR in brief'.

189 See Vedder, 843.

190 Schusterschitz, Stillfried attribute the precedence to the general rules of international law unless the Member States have decided to horizontally agree on concrete measures in accordance to paragraph (2) (see Schusterschitz, Stillfried, (8)). This view is in contrast to the mandatory nature of the whole Article 222 and of paragraph (2). Therefore, when the preconditions for paragraph (2) apply, it is no longer possible to renounce coordination in the Council.

defence policy and by the Committee referred to in Article 71; the two committees shall, if necessary, submit joint opinions."[191]

The provision regulates that, to the extent assistance by the EU is concerned, the arrangements for the implementation of the Solidarity Clause are to be defined by a decision adopted by the Council acting on a joint proposal by the Commission and the High Representative of the Union for Foreign Affairs and Security Policy. The 'Political and Security Committee' (PSC) and the 'Standing Committee on Internal Security' (COSI) assist the Council for these purposes. If necessary, the two committees can submit joint opinions. Also paragraph (3) has been taken over from the implementing provision of the Solidarity Clause in the TCE (Article III-329(2)).

The arrangements for the implementation of the Clause by the Union could be either seen as a general framework covering any relevant situation or "as individual ad hoc acts for every single crisis"[192]. The fact that the Commission and the High Representative have already tabled the proposal for a general framework concerning the arrangements for the implementation of the Solidarity Clause by the Union makes obsolete any discussion whether paragraph (3), first subparagraph, calls for a general framework or *ad hoc* decisions on a case-by-case basis. Moreover, from a practical point of view it would not be appropriate to engage in a lengthy decision-making procedure involving different institutions and bodies. According to the second sentence of paragraph (3) the Council "shall act in accordance with Article 31(1) of the Treaty on European Union where this decision has defence implications". The Joint Proposal presented by the Commission and the High Representative excludes "defence implications" and, therefore, does not provide for a general framework if military means provided by the Member States are used for the purposes of the Clause. Surprisingly, the Commission and the High Representative make the following statement in the explanatory part of the Joint Proposal: "If military support is needed, other than what is already foreseen by the Civil Protection Mechanism, a separate proposal will be submitted by the High

191 Art. 222(3) TFEU.

192 Jacques Keller-Noellet, 'The Solidarity Clause of the Lisbon Treaty's', 332, available at http://www.notre-europe.eu/media/tgae20117fkellernoellet.pdf?pdf=ok.

Representative in accordance with the relevant Treaty provisions."[193] It appears to mean that the Council would adopt decisions on an *ad hoc* basis with regard to the use of military assets. In the absence of a general legal framework the High Representative would have to table *ad hoc* decisions in a timely manner, therefore, the use of military assets beyond what is already covered by the Civil Protection Mechanism would obviously become very unlikely. These decisions would have to be adopted unanimously by the Council, which is not always an easy undertaking.

If the Commission and the High Representative do not also present a general implementing provision for the (maybe unlikely but still possible) case that the Member States provide the Union with military means to assist a Member State, a legal gap will persist and the Solidarity Clause will not be exploited to its full extent. Certainly, the implementing Council Decision could or should have provided for a clearer definition of the use of military capabilities in particular, taking into account the strict limits for the use of armed forces in certain Member States[194].

It is also surprising that, although not part of the CFSP, the Solidarity Clause reflects several characteristics of the CFSP. The legislative procedure for the implementing decision resembles, for instance, similar CFSP procedures where a decision is adopted by the Council based on the joint proposal of the Commission and the High Representative[195]. As no particular reference is made to the voting in the Council, the general rule of Article 16(3) TEU applies according to which the measure has to be decided by qualified majority voting.

The horizontal ('cross-pillar') nature of the Clause is highlighted by the circumstance that the Council may be assisted by the PSC and the COSI, two committees which in their respective policy areas (CFSP for PSC and AFSJ for COSI) enjoy a particular status due to their primary law legal basis. In order to prepare the Council Decision under paragraph (3), both

---

193 18124/12, 5.

194 See Nicolai von Ondarza, Roderick Parkes, ,The EU in the face of disaster' (Stiftung Wissenschaft und Politik, SWP Comments 9 [April 2010]), 7.

195 See Schusterschitz, Stillfried, (13).

committees started discussions on the proper interpretation of the Clause long before the Joint Proposal of the Commission and the High Representative was tabled.

V.a.iv. Article 222(4) TFEU

"4. The European Council shall regularly assess the threats facing the Union in order to enable the Union and its Member States to take effective action."[196]

The last paragraph of Article 222 TFEU introduces a new obligation for the European Council to regularly assess the threats the Union faces in order to enable the Union and its Member States to take effective action. So far the European Council has only dealt with matters of internal and external security on an *ad hoc* basis. The only two 'assessments' of the main challenges with regard to security adopted by the European Council are the 'European Security Strategy' and the 'Internal Security Strategy for the European Union'[197]. However, these two strategies were basically political assessments of the main threats and challenges for the Union. After the adoption of the Internal Security Strategy an attempt was made by the then trio Presidency of the EU Council, Spain, Belgium and Hungary, to provide the Council with a summary of the current threats and challenges to internal security in the EU. This report was called 'First main assessment and description report for internal debate (M.A.D.R.I.D. Report)'[198] and was based on Europol's 'Organised Crime Threat Assessment (OCTA)', Europol's 'Terrorism Situation and Trend Report (TE-SAT)', and the Frontex 'Annual Risk Analysis (ARA)'. The M.A.D.R.I.D. Report aimed at analysing "the specific threats posed by the main internal criminal activities" assessed by the above-mentioned Europol and Frontex "documents [...] as well as the challenges of border control and

---

196 Art. 222(4) TFEU.

197 The Internal Security Strategy (5842/2/10) was adopted by the Justice and Home Affairs Council at its meeting on 25 and 26 February 2010 and was approved on the basis of Article 68 TFEU by the European Council on 25 and 26 March 2010.

198 10203/10.

civil protection. It describes the current situation, trends and impacts, taking into account citizen concerns, and offers conclusions and considerations to tackle these phenomena. Furthermore, it establishes a close link with the proactive approach in the field of disaster management, prevention and all-hazard risk analysis, as was announced in the"[199] Internal Security Strategy. Although the intention to provide the Council with a comprehensive assessment of the main threats in the areas of organised crime, cybercrime, border control and even civil protection was laudable, the actual process of adopting this report was disappointing. Member States tried to negotiate their own priorities into the text of the report. Therefore, the M.A.D.R.I.D. Report was presented to the JHA Council as mere report of the Presidency and so far has remained a unique experiment in providing the Council with a comprehensive (though not complete) assessment of the security situation in Europe.

Article 222(4) TFEU should give new impetus to the establishment of regular assessment for the highest political level in the EU and bring more continuity into the European Council agenda as regards the topics of internal and also external security. This assessment should not only give an overview of the current security challenges but also guidance for the Union and the Member States when taking "effective action". It could be questioned whether this action could also be of a preventive nature and, if yes, for whom: for the Union and for the Member States or only for the former? Could or even should the Union base its preventive action (paragraph (1)(a), first and second indent) on this assessment?

According to Article 8 of the 'Joint Proposal of the Commission and the High Representative on the arrangements for the implementation by the Union of the Solidarity clause' "[t]he Commission and the High Representative will regularly produce, starting in 2015, a joint integrated threat and risk assessment report at Union level. 2. This report shall build on assessments of threats, hazards and risks currently compiled in various sectors (e. g. terrorism, organised crime, civil protection, health, environment, climate change, etc.) based notably on monitoring, interpretation and sharing of information

199 Ibid., 2.

provided by the Member States (through existing sectorial networks or from crisis centres) and Union Agencies as well as relevant international organisations." Paragraph (3) states that "[t]he integrated threat and risk assessment reports will be the basis for a regular assessment by the European Council" [200].

From this provision (which is still awaiting its final adoption and implementation) we can derive the aim of the Commission and the High Representative to provide the European Council with a comprehensive assessment which takes into account not only risk and threat assessments from Union bodies but also from the Member States and relevant international organisations. Nevertheless, we cannot conclude that "effective action" of the Union and the Member States should be exclusively based on this assessment. As the comprehensive assessment for the European Council will probably build upon different sectoral assessments made by Union agencies such as Europol, Frontex and IntCen, among others, the (preventive) action of the Union will most likely rely on these assessments and not on the comprehensive assessment alone.

V.b. Instruments of the Union[201]

Starting with a broad and supranational provision ("the Union and its Members States shall act jointly in a spirit of solidarity"), the Solidarity Clause explicitly refers to the fact that EU institutions should be involved in cooperation, together with the Member States[202]. "The next sentence focuses on the mobilisation of resources, obligating 'the Union' to 'mobilise all the instruments at its disposal, including the military resources made available by

200 Joint Proposal of the Commission and the High Representative on the arrangements for the implementation by the Union of the Solidarity clause by the Union of the Solidarity clause (18124/12), 11.

201 On the measures of the European Union in the fight against international terrorism cf. August Reinisch, 'The Action of the European Union to Combat International Terrorism', in Andrea Bianchi, 'Enforcing International Law Norms Against Terrorism' (Oxford and Portland Oregon [2004]), 119–162.

202 See Sara Myrdal, Mark Rhinard, 6.

the Member States'. Reference to 'all Union instruments' implies EU institutional tools, mechanisms, and resources related to attacks and disasters (much of which represent national resources deployed through European means)."[203]

One of the main instruments which will operate once the Solidarity Clause is triggered will be the Civil Protection Mechanism of the Union (CPM)[204]. The CPM "has been used in response to forest fires and floods in southern EU states as well as in the wake of the Asian tsunami of 2004 and most recently the earthquakes in Haiti and Chile in early 2010"[205]. The CPM shall "facilitate reinforced cooperation between the Community and the Member States in civil protection assistance intervention in the event of major emergencies, or the imminent threat thereof"[206]. According to Article 2 of the Council Decision on the CPM "[t]he Mechanism consists of a series of elements and actions including:

1. the identification of intervention teams and other intervention support available in Member States for assistance intervention in the event of emergencies;
2. the setting-up and implementation of a training programme for intervention teams and other intervention support, and for experts for the teams responsible for assessment and/or coordination (hereinafter assessment and/or coordination teams);
3. workshops, seminars and pilot projects on major aspects of interventions;
4. the establishment and dispatch of assessment and/or coordination teams;
5. the establishment and management of a Monitoring and Information Centre (MIC), which is accessible and able to react immediately 24 hours a day and serving the Member States and the Commission for the purposes of the Mechanism;

---

203 Ibid.

204 Council Decision 2007/779/EC, Euratom of 8 November 2007 establishing a Community Civil Protection Mechanism (recast) [2007] OJ L 314/9.

205 Ondarza, Parkes, 2.

206 'Council Decision establishing a Community Civil Protection Mechanism (recast)', 2007/779/EC, Euratom, Article 1(1).

6. the establishment and management of a Common Emergency Communication and Information System (CECIS) to enable communication and sharing of information between the MIC and the contact points of the Member States; [...]
7. supporting consular assistance to EU citizens in major emergencies in third countries regarding civil protection activities if requested by the consular authorities of the Member States;
8. other supporting and complementary action necessary in the framework of the Mechanism as mentioned in Article 4 of Council Decision 2007/162/EC, Euratom of 5 March 2007 establishing a Civil Protection Financial Instrument."[207]

The Community Civil Protection Mechanism and the Civil Protection Financial Instrument together cover three of the main aspects of the disaster management cycle, which is divided into 'prevention', 'preparedness' and 'response'. Participating States in the CPM are all 28 EU Member States and the EFTA states[208]. The CPM also covers the use of military assets. According to Article 4(5) of the CPM Council Decision Member States "may, subject to appropriate security constraints, provide information about relevant military assets and capabilities that could be used as a last resort as part of the civil protection assistance through the Mechanism, such as transport and logistical or medical support". Certainly according to Article 8(7), subparagraph (5), "[i]n the event of a major emergency occurring outside the Community, the possible use of military assets and capabilities available to support civil protection should be consistent with the principles of relevant United Nations Guidelines". Although the Solidarity Clause is not part of the CSDP, it does not refer to any particular instrument of assistance except to "military resources made available by Member States". This "suggests a mobilisation of military assets, either such assets that have already been registered for use in civilian disasters, or a much wider range of resources including, for example, troops for crowd

207 Ibid., Article 2.

208 See European Commission, 'Civil protection legal texts', available at http://ec.europa.eu/echo/civil_protection/civil/prote/legal_texts.htm

control or equipments for large scale disaster clean up"[209]. Whether the use of military means is meant to be of a police nature[210] can be derived neither from the text nor from the preparatory work. The use of military assets in the framework of the CPM is covered by the Joint Proposal. The use of military means going beyond the CPM would require a separate provision in accordance with Article 31(1) TEU.

V.c. Legal Status of the Solidarity Clause

The Solidarity Clause is under the jurisdiction of the Court of Justice of the European Union. Article 275 TFEU only limits the jurisdiction of the CJEU with regard to the CFSP and with respect to acts adopted on the basis of the CFSP provisions[211]. Therefore, as regards the defence provisions the application of the Solidarity Clause "does not wholly lie within the jurisdiction"[212] of the CJEU. Moreover, the jurisdiction of the CJEU is also limited with respect to the validity or proportionality of operations carried out by the police or other law-enforcement services of a Member State (Article 276, TFEU). According to Fuchs-Drapier this "means that most of the police and military operations carried out under the application of the Solidarity Clause following a terrorist attack cannot be controlled"[213] by the CJEU. The fact that the Clause is under the jurisdiction of the CJEU indicates that it is likely to be

---

209 Ibid.

210 Cf. Chinchón Alvarez, J. y Sánchez-Bayón, 'Comentario a los artículos I-43, III-284 y III-329 del Tratado por el que se establece una Constitución para Europa. Los artículos I-43 y III-329: Demasiadas puertas abiertas', in Brú Purón, C. M. (director), 'Exégesis Conjunta de los Tratados Vigentes y Constitucional Europeos' (Thomson-Civitas, Navarra, [2005]), 171: „Pudiera concebirse, no obstante, una interpretación que, sin estar ligada a la idea del combate del terrorismo, habría de situarnos en un escenario de utilización de esos medios militares en tareas meramente policiales, en relación a la prevención y protección, o asistenciales, en caso de ataques consumados."

211 See Schusterschitz, Stillfried, (6).

212 Fuchs-Drapier, 186.

213 Ibid.

enforceable as a Treaty provision[214]. Certainly, "even if the inapplicability of the Clause could be invoked" before the CJEU 'on grounds of lack of competence, infringement of an essential procedural requirement, infringement of the Treaties or of any rule of law relating to its application' (Articles 277 and 263(2) TFEU), "the conditions under which any party can bring the matter before the"[215] Court are very restrictive.

Although the Council Decision on the basis of paragraph (3) has not yet been adopted, a Member State affected by a crisis situation falling under Article 222 TFEU could ask for assistance. It is, however, debatable whether the affected Member State could only expect assistance from the Member States (as no implementing provision is requested) or whether the Union could already assist as well and, if yes, which kind of assistance the Union could provide. According to Fuchs-Drapier "[a]s long as the arrangements for its implementation are not defined, the Solidarity Clause has no direct effect"[216].

The distinction as to whether the character of the Clause is that of a political principle, legal principle or legal rule[217] is also relevant for its enforcement and applicability[218]. The issue of enforceability of the Clause raises a number of difficult questions as Fuchs-Drapier outlines: "The issues raised by the application of the Solidarity Clause seem more sensitive in case of a terrorist attack than in case of a manmade or a natural disaster. The application of the Clause during such an event might indeed hamper democratic control, citizens' rights, national secrecy, and constitutional and judicial traditions. In addition, the executive employment of military forces appears more likely in the event of a terrorist attack. The executive use of military forces is innately different from the support provided by armed forces in the event a natural or unintentionally caused manmade disaster. In Germany, for instance, it requires a declared state

214 See Shev, 26.
215 Fuchs-Drapier, 186.
216 Ibid.
217 On the distinction concerning the character of the solidarity clause as being a political principle, legal principle or legal rule cf. Shev, 20.
218 See ibid., 20.

of defence according to the Basic Law, only possible with a two-thirds parliamentary majority."[219]

The Member States have 'applied' the Solidarity Clause only once, namely after the Madrid bombings of 11 March 2004. After this 'political' enforcement of the Clause the Member States were reluctant to "activate binding procedures that may contest the sovereignty principle since the apparent risk of a terrorist attack on EU soil has tapered off"[220]. The legal relevance of the Clause only "rests on the hypothesis that a Member State would request assistance through the Solidarity Clause in the event of a terrorist attack. *A priori*, few Member States want to entrust the response phase to the Union's responsibility. But the possibility that a small Member State may do so cannot be totally excluded"[221]. If the Solidarity Clause is invoked and a legal dispute on the correct application of the Clause arises, the CJEU could take on new and far reaching tasks, particularly if it takes the opportunity to interpret the CFSP-oriented parts of the Clause[222]. However, the means for the Commission and the CJEU to foster further integration *via* the Solidarity Clause appears to have practical constraints. Indeed, it may prove particularly difficult to assess whether a Member State has complied with the obligation to assist if it is not the affected Member State who initiates infringement proceedings[223].

219 Fuchs-Drapier, 185.

220 Ibid., 184.

221 Ibid., 185.

222 See Schusterschitz, Stillfried, (6).

223 See Hafner, ‚Beistands- und Solidaritätsklausel', 403.

VI. Joint Proposal on the Implementation by the Union of the Solidarity Clause

On 21 December 2012 the High Representative of the EU for Foreign Affairs and Security Policy and the European Commission issued their 'Joint proposal for a Council Decision on the arrangements for the implementation by the Union of the Solidarity Clause'[224] which aims to comply with paragraph (3) of Article 222. This provision of Article 222 TFEU requires the Commission and the High Representative to submit to the Council a proposal on the Union's arrangements for implementing the Clause.

On the initiative of the Hungarian Presidency of the EU Council in the first half of 2011 and in accordance with the first 18-month programme of the Salzburg Forum[225], the Article 71 TFEU Committee (COSI) started to first informally[226] and then (under the Polish Presidency in the second half of that year) to formally discuss questions concerning the proper implementation of Article 222(3) TFEU. We may recall that initially the Commission and the High Representative had already planned to present a proposal on a draft

---

224 Joint proposal for a Council Decision on the arrangements for the implementation by the Union of the Solidarity Clause, JOIN(2012) 39 final (18124/12).

225 The Salzburg Forum, based on an Austrian initiative taken in 2000, is a platform for multilateral dialogue and cooperation on issues of internal security. Its member states are Austria, Bulgaria, Croatia, the Czech Republic, Hungary, Poland, Romania, Slovakia and Slovenia. The principal idea of the founding states of the Salzburg Forum in 2000 was to support the European integration of Austria's Eastern neighbours and to facilitate their efforts towards EU accession. In the meantime, all Salzburg Forum states have become EU Member States, sharing common security interests and standards (http://www.salzburgforum.org/site/aboutus.html).

226 In the second meeting of the 'Strategic Circle Internal Security' (founded on the initiative of Austria on 1 November 2010 with the aim to discuss strategic issues of internal security policy among different Member States represented by their senior representatives in COSI and/or CATS), which took place on the 14 and 15 January 2011, Austria asked for the triggering of a debate on the implementation of the Solidarity Clause.

Council Decision implementing the Solidarity Clause in 2010[227]. The first formal debate was held by CATS[228] in its meeting on 6 October 2011[229] in order to give guidance for the preparation of a broader debate which took place at a meeting of the members of the Political and Security Committee (PSC) and of the members of the Standing Committee on operational cooperation on internal security (COSI) on 14 December 2011.

VI.a. Content

In the explanatory part of the Joint Proposal, the Commission and the High Representative stress the fact that "[d]ue to the wide scope of the Treaty article, the arrangements for implementing the Solidarity Clause cover a wide number of policy areas and instruments"[230]. According to the authors of the Joint Proposal, the policy areas affected by it are "the EU Internal Security Strategy[231], the Union Civil Protection Mechanism and the Civil Protection Financial Instrument[232], the EU Solidarity Fund[233]; the health security initiative for serious cross-border threats to health[234], crisis response and analysis structures in the EEAS, and the Crisis Coordination Arrangements in the

---

227 See Schusterschitz, Stillfried, ‚Kommentar zu EUV und AEUV', ‚Art 222', (13).

228 The 'Coordinating Committee in the area of police and judicial cooperation in criminal matters' (ex-'Article 36 Committee' or 'CATS') was established by COREPER (16070/09).

229 See 'Solidarity Clause – results of the orientation debate on Art. 222 TFEU at the CATS meeting on 6 October 2011' (17185/11).

230 18124/12, 3.

231 Communication from the Commission to the European Parliament and the Council of 22 November 2010–The EU Internal Security Strategy in Action: Five steps towards a more secure Europe (COM(2010) 673 final).

232 Council Decision establishing a Community Civil Protection Mechanism (recast), 2007/779/EC, Euratom; Council Decision establishing a Civil Protection Financial Instrument, 2007/162/EC, Euratom.

233 Council Regulation (EC) No 2012/2002 of 11 November 2002 establishing the European Union Solidarity Fund.

234 Proposal for a decision of the European Parliament and of the Council on serious cross-border threats to health, COM(2011) 866 final.

Council"[235]. Moreover, the "arrangements are also coherent with the establishment of a European Area of Justice in the Union"[236].

As stated in its explanatory memorandum, "[t]he proposal defines the geographic scope, the activation mechanism, and the response arrangements at Union level"[237]. The Joint Proposal also takes into consideration and is "consistent with the arrangements for coordination in the Council (on the basis of the Crisis Coordination Arrangements), in accordance with Article 222(2)"[238]. Moreover, the Joint Proposal by recalling paragraph (4) of Article 222 "sets out arrangements for: (i) an integrated threat and risk assessment at Union level to be used as a basis for a regular assessment by the European Council and (ii) preparedness measures by the Union and Member States, on the basis of the guidance of the European Council"[239]. All in all, the proposal of the High Representative and the Commission is meant to be a practically comprehensive instrument for the implementation of Article 222 as it also encompasses the 'intergovernmental'[240] paragraph (2) and paragraph (4).

Although the Joint Proposal focuses on the Union's implementation of the Solidarity Clause, according to the first recital of the draft Council Decision also "coherence and complementarity of Union action should be sought with other assistance to be provided by Member States pursuant to Article 222(2) of the Treaty and Declaration 37 annexed to the Treaty, which lays down that a Member State can choose the most appropriate means to comply with its own solidarity obligation towards another Member State"[241]. During discussions in CATS and PSC/COSI several Member States asked for an implementation of the Solidarity Clause that should avoid duplication of work and rely on existing instruments, a fact which is highlighted in recital (2) of the draft Council Decision. A further recital is worth mentioning:

---

235 18124/12, 3.
236 Ibid.
237 Ibid., 2.
238 Ibid., 3.
239 Ibid.
240 Cf. Vedder, 843.
241 18124/12, 7.

> "(12) Where necessary and practicable in view of the urgency, the response arrangements at Union level should be complemented by the adoption of legal acts or the amendment of existing acts in accordance with the relevant provisions of the Treaty."

In recital (12), which is more or less[242] equivalent to Article 5(1)(d) of the Joint Proposal, we can clearly recognize the aim of the authors of the Joint Proposal to emphasize the integrationist potential of the Clause. However, the *ad hoc* establishment or revision of legal provisions under the impression of individual calamitous events in an area (i.e. civil protection) where the Union merely has supporting legislative competence should be avoided.

According to recital (13), the "decision has no defence implications. In case a crisis requires Common Foreign and Security Policy (CFSP) action, beyond the use of military resources covered by existing arrangements on civil protection, a decision should be taken by the Council in accordance with relevant Treaty provisions."

Looking at the legal text we find the scope of the Council Decision in Article 2:

> "This Decision shall apply in case of terrorist attacks or natural or man-made disasters, irrespective of whether they originate inside or outside the territory of the Member States:
> (a) Within the territory of Member States to which the Treaty applies, including land area, territorial sea and airspace.
> (b) When affecting ships (when in international waters) or airplanes (when in international airspace) or critical infrastructure (such as off-shore oil and gas installations) when under the jurisdiction of a Member State."

242 Recital (12) explicitly refers to the adoption of new or amendment of existing legal acts in accordance with the appropriate provisions in the Treaties, therefore, also in accordance with the ordinary legislative procedure. Article 5(1)(d) only mentions the Council as addressee of the proposals of the Commission and the High Representative.

Article 222 TFEU refers to the 'territory' of Member States. Its wording on the protection of democratic institutions and the civilian population, however, does not contain any territorial limitation. Nevertheless, ships, airplanes or critical infrastructure are not to be considered the territory of the home State according to international law. Moreover, they do neither represent democratic institutions, nor are they necessarily connected to the civilian population (the civilian population of a Member State could, for example, also be in a hotel resort in a third State). Therefore, the Joint Proposal on the one hand extends the geographical scope of Article 222 TFEU to ships, airplanes or critical infrastructure, and is also in contrast to international law. The question of geographical scope has been debated during the negotiations on the Joint Proposal in the second half of 2013 and was also the object of a written opinion[243] of the Council Legal Service (CLS). The CLS comes to the conclusion that Article 2(b) of the draft Council Decision does not infringe international law[244].

In its Article 3 ('Definitions') the draft Council Decision introduces the term 'crisis' which is not mentioned in Art. 222 TFEU. According to Article 3(a) a crisis is "a serious, unexpected and often dangerous situation, requiring timely action; a situation that may affect or threaten lives, environment, critical infrastructure or core societal functions, may be caused by a natural or man-made disaster or terrorist attacks". Under paragraph (b) the draft Council Decision specifies the notion of disaster: "Disaster: any situation, which has or may have an adverse impact on people, the environment or property." It is questionable whether a differentiation between 'crisis' and 'disaster' is necessary or even advisable. As regards the requirement of unexpectedness, it appears relevant to establish if, for example, in cases of gross negligence by the affected Member State the Union and the Member States remain bound by the Solidarity Clause (and, therefore, could theoretically incur infringement

243 Contribution of the Council Legal Service: Joint proposal for a Council Decision on the arrangements for the implementation by the Union of the Solidarity Clause (Article 222 TFEU) - Scope (14498/13).

244 Despite the written contribution of the CLS by the end of 2013, the question on the geographical scope remained controversial.

proceedings if they do not act 'in a spirit of solidarity'). Moreover, the notion of "disaster" according to the Joint Proposal is much too broad if the Solidarity Clause is to be an instrument of last resort (the explanatory memorandum says that "the EU should act only in exceptional circumstances"[245]).

There is no clear definition of what falls under the Joint Proposal. Any situation having an "adverse impact" seems to indicate a very low threshold requirement. The Commission and the High Representative referred to the same definition of disaster covered by the new 'Commission Proposal for a Decision of the European Parliament and of the Council on a Union Civil Protection Mechanism'[246]. However, during the negotiations of the new CPM proposal Member States were already reluctant to accept such a vague definition of disaster and asked for amendments underlining the necessity of a higher threshold requirement. There is a concrete risk that, under a triggering system combined with the definition of disaster from the Joint Proposal, "well-equipped member states would probably make little use of the solidarity clause, whilst accident-prone states could rely upon it at the expense of the maintenance of their own capabilities"[247]. It remains uncertain whether the possible solution to this problem could be found in setting a 'subsidiarity baseline' which should work as a "second lock on EU action"[248]. The idea of establishing beforehand a "level of severity under which the affected member state would be obliged to deal with disasters themselves"[249] is in principle reasonable but would in practice confine the Clause to the destiny of a 'dead letter provision'. Member States would probably avoid any risk that their request for assistance in accordance with Article 222 TFEU could be denied due to non-compliance with a certain level of seriousness of the incident in question. In principle, it will be up to the affected Member State to make a first assessment of the kind of assistance needed and table a request on the political level to the other Member States. In order to have legal certainty as to whether the relevant incident was actually

245 18124/12, 3.

246 COM(2011) 934 final.

247 Ondarza, Parkes, 3.

248 Ibid., 4.

249 Ibid.

a case suited to invoking the Solidarity Clause with all its consequences (also with regard to possible infringement proceedings), a declaration of the Council confirming among other things the correctness of the application of Article 222 TFEU would probably be the better solution.

The Joint Proposal sets out the procedure to be followed for the activation of the Solidarity Clause. Article 4 established that a "Member State being the object of an actual or imminent terrorist attack or the victim of a natural or man-made disaster may invoke the Solidarity Clause if, after having exploited the possibilities offered by existing means and tools, at national or Union level, it considers that the situation overwhelms its response capacity" (paragraph (1)). "The Member State affected shall address its request to the President of the European Commission, through the Emergency Response Centre, and notify simultaneously the Presidency of the Council" (paragraph (2)). The Emergency Response Centre which was later renamed into Emergency Response Coordination Centre – ERCC) acts as the initial, single, 24/7 point of contact for the competent authorities of the Member State affected. The Emergency Response Coordination Centre (ERCC) was launched in May 2013 and replaces and upgrades the functions of the previous Monitoring and Information Centre (MIC)[250]. ERCC can "deal with several simultaneous emergencies in different time zones, around-the-clock"[251]. It acts as a "coordination hub facilitating a coherent European response during emergencies helping to cut unnecessary and expensive duplication of efforts. It collects and analyses real-time information on disasters, monitors hazards, prepares plans for the deployment of experts, teams and equipment, and works with Member States to map

250 The Monitoring and Information Centre (MIC) was the 'operational heart' of the CPM. It was based at the European Commission in Brussels and accessible 24/7. It could spring into action immediately when it received a call for assistance. The MIC worked in close cooperation with national crisis centres throughout the 32 countries participating in the Mechanism (EU 28, the Former Yugoslav Republic of Macedonia, Iceland, Liechtenstein and Norway) (see the information of the European Commission on the Monitoring and Information Centre at http://ec.europa.eu/echo/policies/disaster_response/mic_en.htm).

251 European Commission, Emergency Response Coordination Centre (ERCC), available at http://ec.europa.eu/echo/about/ERC_en.htm.

available assets and coordinate the EU's disaster response efforts by matching offers of assistance to the needs of the disaster-stricken country."[252]

At the core of the Joint Proposal are the response arrangements at Union level under Article 5:

"1. Once the Solidarity Clause has been invoked, the Commission, and the High Representative in accordance with article 1.2, shall:

(a) identify and use all relevant Union instruments that can best contribute to a response to the crisis, including sector-specific, operational, policy or financial decisions (e.g. the Civil Protection Mechanism, the Strategic Analysis and Response Centre, the Health Emergencies Operations Facility, the intelligence resources of IntCen) that fall in the remit of the Commission and of the High Representative, as well as military resources mobilised through EUMS; in addition, identify and propose the use of instruments and resources falling within the remit of Union Agencies;

(b) assess whether existing instruments are sufficient;

(c) produce regular integrated situation assessment and analysis reports to inform and support the coordination and decision-making at political level in the Council;

(d) where appropriate, submit proposals to Council, in particular as concerns: operational decisions on reinforcement of existing mechanisms, decisions on exceptional measures from Member States not foreseen by existing instruments; policy coordination and information exchange with the aim to create necessary regulatory conditions; operational or support measures for fast reaction of Member States.

The Commission shall convene meetings to prepare the proposed crisis response measures. The Commission shall invite the EEAS and the relevant EU Agencies.

2. The Emergency Response Centre (ERC) shall initially act as the single operational hub with Member States at Union level. The Commission, in

252 Ibid.

consultation with the High Representative, may subsequently designate another centre better placed to assume that function taking into account the nature of the crisis. The designated operational hub shall act as the primary entry point for Member States. It shall lead operational response coordination and produce situation assessment reports."[253]

The Joint Proposal aims to set up a system where the Commission and the High Representative are the sole key players in the response to a crisis. The Council has no active role and the European Parliament is not even mentioned. With regard to the latter it should be noted that the European Parliament adopted a resolution on the 'EU's mutual defence and solidarity clauses'[254] where it asked for the "proper involvement of the national parliaments and of the European Parliament" (paragraph (43)) and noted "that the European Parliament and the Council, as the EU legislators and budgetary authorities, should be kept informed of the situation 'on the ground' in the case of a disaster or attack that triggers the solidarity clause, as well as of its origins and possible consequences, so that a thorough and unbiased assessment based on up-to-date and concrete information can be carried out for future reference" (paragraph (44)).

The fact that the authors of the Joint Proposal established such a Commission-driven system deserves to be thoroughly analysed: first of all, it is to be noted that Article 222 TFEU does not explicitly foresee such a prominent role of the Commission and the High Representative. According to Article 222(3) TFEU, the Council is assisted by PSC and COSI. This would rather be an indication that the Council decides or at least has a central role in the decision on the assistance to be rendered by the Union. There is also certain incongruence with regard to Article 4 of the Joint Proposal, which establishes that the activation of the Clause can only take place after the affected Member State has exploited the possibilities offered by existing means and tools at national or Union level. In Article 5 it is the Commission and the High Representative

253 18124/12, 10.

254 European Parliament resolution of 22 November 2012 on the EU's mutual defence and solidarity clauses: political and operational dimensions (2012/2223(INI)).

who identify and use all relevant EU instruments after the Clause has been invoked. The question arises as to what further "possibilities offered by existing means and tools, at national or Union level" exist that the affected Member State has not exploited but could still be identified and used by the Commission and the High Representative. Does this mean that, due to the complexity of the situation, the existing means and tools at national and Union level would not suffice if employed independently from each other and that a number of different EU tools would have to be used in combination? If this is the correct interpretation, it should be clearly spelt out in the text of the decision itself. In the event of possible financial assistance from the EU Solidarity Fund, the relevant provisions (thresholds) to activate the fund should apply. A further question arises, namely what is meant by "operational decisions"[255] as there is no such term in the TFEU or TEU. Are operational tasks conferred upon the Commission and the High Representative? With regard to the provision in Article 5(2) of the Joint Proposal, it seems questionable whether it makes sense to automatically task the ERCC as the initial hub in every case, even in cases where the origin of the disaster is clearly a terrorist attack. In such a situation Europol (if necessary, supported by IntCen) should probably be tasked from the beginning as a "single operational hub".

The Council is only called into play by the Presidency on an optional basis. Article 6 of the Joint Proposal stipulates that "[u]pon activation of the Clause, the Presidency of the Council may take the decision to activate the Crisis Coordination Arrangements and identify the most appropriate way to prepare rapid consultations and decisions in the Council, in respect of the obligation to assist. Support to the operation of the Crisis Coordination Arrangements shall be provided by the General Secretariat of the Council, the Commission and EEAS." The role of the Council is confined to the Crisis Coordination Arrangements (CCA) (now: Integrated Political Crisis Response Arrangements[256] [IPCR]). On the one hand this approach makes sense as no new structures need to be created; on the other hand it is not obvious that

255 Article 5(1)(d) of the Joint Proposal.

256 See the document 'Finalisation of the CCA review process: the EU Integrated Political Crisis Response (IPCR) arrangements' (10708/13).

political coordination in the framework of the Clause always requires the activation of the Crisis Coordination Arrangements (CCA/IPCR). However, the new Integrated Political Crisis Response Arrangements should be flexible enough to guarantee an appropriate political and/or operational response of the Council and the Member States.

Articles 7 and 8 are dedicated to the implementation of Article 222(4) TFEU. Article 7 foresees the compilation of 'situation assessment reports' which will be "compiled by the ERC or the designated operational centre, in collaboration with the EU Situation Room. The preparation of such reports will draw on contributions from the different situation awareness and crisis management centres in Member States, in the Commission, the EEAS and the relevant EU agencies, as well as in relevant International Organisations"[257]. Article 8[258] introduces so-called 'Integrated threat and risk assessment at Union level' which will be first realised in 2015. This 'Integrated threat and risk assessment at Union level' is to be the basis for the assessment by the European Council as covered in 222(4) TFEU. It is noteworthy that there is no mention of 'risk' assessment but only of 'threat' assessment in Article 222 TFEU. Threat assessments focus more on the causes of vulnerability and dangers, whereas risk assessments address the probability of suffering damage, loss or destruction. Therefore, the Joint Proposal extends the scope of Article 222(4) TFEU and

257 Article 7 Joint proposal for a Council Decision on the arrangements for the implementation by the Union of the Solidarity Clause.

258 Article 8 Joint proposal for a Council Decision on the arrangements for the implementation by the Union of the Solidarity Clause:

"1. The Commission and the High Representative will regularly produce, starting in 2015, a joint integrated threat and risk assessment report at Union level.
2. This report shall build on assessments of threats, hazards and risks currently compiled in various sectors (e.g. terrorism, organised crime, civil protection, health, environment, climate change, etc.) based notably on monitoring, interpretation and sharing of information provided by the Member States (through existing sectorial networks or from crisis centres) and Union Agencies as well as relevant international organisations.
3. The integrated threat and risk assessment reports will be the basis for a regular assessment by the European Council.

introduces a potentially complex process involving diverse actors such as EU agencies, international organizations and Member States. The production of such assessments will therefore require thorough preparation and proper methodology in order to avoid discussions on the assessment itself.

VI.b. Recent developments

The Joint Proposal is currently being discussed in a 'Friends of the Presidency' group within the framework of the Council. The mandate of the group was adopted by COREPER in February 2013[259] and builds upon the Friends of Presidency Group which dealt with the review of the CCA. Little progress on the text of the Joint Proposal was achieved during the first half of 2013. This was mostly due to the fact that most Member States were asking for a stronger involvement of the Council, whereas the Commission and the EEAS showed little understanding for the arguments of the Member States. The adoption of the Joint Proposal, therefore, which was originally foreseen for December 2013 will happen in 2014.

259 See the mandate in Council document 'Handling within the Council of the joint proposal by the Commission and the High Representative on the arrangements for the implementation by the Union of the Solidarity Clause (Art. 222 TFUE)' (6598/13).

VII. Relation between Article 42(7) TEU and Article 222 TFEU

The Mutual Assistance Clause foresees that, if a Member State suffers armed aggression on its territory, its partners have an obligation to provide aid and assistance "by all their means in their power", in accordance with Article 51 of the UN Charter. Article 42(7) TEU introduces an intergovernmental obligation for the Member States to provide aid and assistance on a national basis. It is, in the end, the individual Member State who decides how to provide assistance[260]. The Solidarity Clause is supranational in its remit and brings in a common obligation of the Union and the Member States. Contrary to the Solidarity Clause, the Mutual Assistance Clause only applies after armed aggression has taken place. No preventive elements can be derived from Article 42(7) TEU. The relationship between both clauses becomes, therefore, only relevant when a terrorist attack has taken place on the territory of a Member State.

There appear to be both clear differences and confusing similarities between the two Clauses. Differences are, among other things, that Article 42(7) applies to situations of 'armed aggression' against the territory of a Member State and is, therefore, geographically more limited than the Solidarity Clause. The Mutual Assistance Clause is purely intergovernmental in character and does not confer any role on the Union. An activation of Article 42(7) TEU does not require political coordination at the EU level[261]. "These differences should not obscure similarities between the two clauses, however. Both introduce binding commitments amongst Member States. Both are predicated on the ability to draw on 'all available means' when requested. There are theoretical possibilities that both clauses could be triggered together, particularly in cases when the 'threat agent' is unclear (the Estonian cyber attacks, for

260 See Bruno Angelet, Ioannis Vrailas, 'European Defence in the wake of the Lisbon Treaty', (Egmont Royal Institute, Egmont Paper 21 [May 2008]), 20.

261 See ibid., 9.

example)."[262] With regard to cyber attacks, it could indeed be difficult to draw an exact line between the two Clauses. The question of whether cyber attacks can be qualified as an armed attack (aggression) is still being debated. Some scholars appear to focus on the consequences of a cyber attack and whether the effects of cyber attacks are tantamount to a military attack[263]. A legally very restrictive view of cyber attacks suggests that in order to qualify a cyber attack as armed attack it must produce "*violent* consequences"[264]. All in all, it appears that cyber attacks could be qualified as armed aggression (attack). Likewise, cyber attacks could be interpreted as terrorist attacks. "Concepts such as 'armed aggression' are not static but evolve together with a changing security landscape. This creates grey zones in defining the applicability of the mutual defence clause in situations involving large-scale attacks on Member States' information and communication networks."[265]

As already outlined above, possible overlaps could arise if a terrorist attack committed by non-State entities affects a Member State on its territory[266]. With respect to terrorist attacks the Solidarity Clause would result as *lex specialis* but would not be applicable in the event of a terrorist attack perpetrated by a third State[267]. As Konstadinides observes, no obstacle in the TEU "would have prevented CSDP actions entailing the use of Member States' armed forces

---

262 Ibid., 9.

263 See Abraham D. Sofaer, David Clark, Whitfield Diffie, 'Cyber Security and International Agreements', in 'Proceedings of a workshop on deterring Cyberattacks: Informing strategies and developing options for U.S. policy (2010)', 185, available at http://www.nap.edu/openbook.php?record_id=12997&page=185. ("[T]he right of states to exercise self-defense or to take countermeasures in response to such attacks would depend on their potential consequences.").

264 Yoram Dinstein, 'Computer Network Attacks and Self-Defense' (76 INT'L L. STUD. 99 [2002]), 103. Cited by Matthew C. Waxman, 'Cyber-Attacks and the Use of Force: Back to the Future of Article 2(4)' (THE YALE JOURNAL OF INTERNATIONAL LAW [Vol. 36]), 235, available at http://www.yjil.org/docs/pub/36–2-waxman-cyber-attacks-and-the-use-of-force.pdf.

265 Myrdal, Rhinard, 11.

266 See Hafner, ‚Beistands- und Solidaritätsklausel', 382–383.

267 See Calliess, 2189.

to realise soft security objectives within the EU in times of emergency"[268]. Certainly, "a clearer demarcation of pillar-overarching areas would be desirable so that the competence under CFSP and TFEU would occupy different fields"[269]. According to Konstadinides "Article 42 (7) TEU could deal with state violence against one or more Member States and statesponsored terrorism through CSDP military means. On the other hand, Article 222 TFEU could focus on countering oppositional terrorism by non-state actors and serious crime through the EU Criminal law *acquis.*"[270] Be that as it may, where the threat or the origin of the terrorist attack is ambiguous, a parallel activation of both Clauses will probably be inevitable[271].

All in all, a more in-depth analysis on the part of the EU institutions regarding the scope of both Clauses would be necessary. Unfortunately, during the negotiations of the Joint Proposal the European Commission, the EEAS and the Council Legal Service have been reluctant to provide a thorough legal analysis of several legal questions including the possible overlaps of both Clauses. Providing for legal certainty as far as possible would encourage Member States to make use of these instruments and could spare the Clause from a 'dead letter' fate where Member States would first and foremost resort to bi- or multilateral instruments.

268 Theodore Konstadinides, 'Evidence on the Government's Review of the Balance of Competences between the United Kingdom and the European Union', 8, available at https://www.gov.uk/government/uploads/system/uploads/attachment_data/file/224408/evidence-theodore-konstadinides.pdf.

269 Ibid., 9.

270 Ibid.

271 See ibid., 11.

VIII. Relation to other Provisions in the Treaties

The Mutual Assistance and the Solidarity Clauses are not the only provisions in the EU treaties that cover crisis situations. The Union has a wide array of instruments to deal with exceptional occurrences in a spirit of solidarity, such as the Civil Protection Mechanism, the Solidarity Fund[272], and the possibility to grant economic and financial support in cases of severe difficulties, as provided for in Article 122 TFEU. Moreover, the Solidarity Clause touches upon the newly established legal basis for the policy area 'Civil Protection' as well as the provision concerning 'Humanitarian Aid' in Article 214 TFEU, whereas the relation of both Clauses with regard to Articles 346 and 347 TFEU is particularly relevant for the neutral and non-aligned Member States.

VIII.a. Relation to Article 122 TFEU[273]

Article 122 TFEU opens the possibility for the Council to adopt "in a spirit of solidarity between Member States" appropriate measures "in particular if severe difficulties arise in the supply of certain products, notably in the area of energy" (paragraph (1)). Article 122 TFEU "safeguards intergovernmental decision by stating that the Council 'may' decide appropriate measures and grant financial

---

272 Council Regulation (EC) 2012/2002 of 11 November 2002 establishing the European Union Solidarity Fund [2002] OJ L 311/3.

273 Article 122 TFEU:

"1. Without prejudice to any other procedures provided for in the Treaties, the Council, on a proposal from the Commission, may decide, in a spirit of solidarity between Member States, upon the measures appropriate to the economic situation, in particular if severe difficulties arise in the supply of certain products, notably in the area of energy.

2. Where a Member State is in difficulties or is seriously threatened with severe difficulties caused by natural disasters or exceptional occurrences beyond its control, the Council, on a proposal from the Commission, may grant, under certain conditions, Union financial assistance to the Member State concerned. The President of the Council shall inform the European Parliament of the decision taken."

assistance according to each situation"[274]. Financial assistance can be granted to a Member State which "is in difficulties or is seriously threatened with severe difficulties caused by natural disasters or exceptional occurrences" (paragraph (2)). The notion of 'exceptional occurrence' could encompass situations that are no longer manageable by the affected Member State, such as social unrest, but also the involvement in military conflicts[275]. Due to its scope, Article 122 TFEU could obviously overlap with Art. 222 TFEU. If a Member State is affected by a natural disaster (or man-made disaster), financial assistance could eventually also be a measure taken in accordance with the Solidarity Clause. It should be noted that Article 122 only applies after a natural disaster has taken place and within the territory of the Member States[276]. The wide spectrum of applicability of this provision can be shown by the fact that it was also used as the legal basis for setting up the European Financial Stabilisation Mechanism (EFSM)[277].

In order to decide which provision should prevail we should consider the text of Article 122(1). The provision states that the Council may take appropriate measures "[w]ithout prejudice to any other procedures provided for in the Treaties". This suggests more specific provisions should not be replaced by Article 122 TFEU. In this respect, consideration should be given to the jurisdiction of the CJEU with regard to the 'democratic principle' referring to the precedence of legislative procedure where the European Parliament is involved (see Roquette Frères vs. Council[278] and Maizena vs. Council[279])[280]. Nevertheless, neither Article 122 TFEU nor Article 222 TFEU confers any role on the European Parliament in their legislative procedures. In both cases the Parliament is only to

274 Fuchs-Drapier, 185.

275 See Michael Rodi, ‚Artikel III-180‘, in Christoph Vedder, Wolff Heintschel von Heinegg, ‚Europäischer Verfassungsvertrag‘ (Baden-Baden: Nomos Verlag [2007]), 528.

276 See Schusterschitz, Stillfried, (6).

277 Council Regulation (EU) No 407/2010 of 11 May 2010 establishing a European financial stabilization mechanism [2010] OJ L 118/1.

278 Roquette Frères vs. Council (Case 138/79) [1980] ECR 3333.

279 Maizena vs. Council (Case 139/79) [1980] ECR 3393.

280 See Peter Part, ‚Artikel 120–126 AEUV‘, in Heinz Mayer, Karl Stöger, ‚Kommentar zu EUV und AEUV unter Berücksichtigung der österreichischen Judikatur und Literatur‘, (Wien: Manz Verlag [2012]), (30).

be informed. Therefore, no precedence to the one or to the other provision can be derived from the respective legislative procedure. Probably both provisions could be applied cumulatively as regards financial assistance after natural disasters and exceptional occurrences falling under the remit of both provisions. As regards severe financial and economic crisis situations, Article 122 TFEU would obviously be the more suitable legal basis[281]. However, we have to consider that according to Article 4(1) of the Joint Proposal a "Member State being the object of an actual or imminent terrorist attack or the victim of a natural or man-made disaster may invoke the Solidarity Clause if, after having exploited the possibilities offered by existing means and tools, at national or Union level, it considers that the situation overwhelms its response capacity". It could be questioned whether Article 122 TFEU falls under the scope of this provision. In this case the affected Member State would have to invoke Article 122 TFEU first and only if this provision cannot be 'exploited' does the Solidarity Clause come into play.

VIII.b. Relation to Article 196 TFEU[282]

The Treaty of Lisbon introduces a supporting competence in the policy area of civil protection under Article 6 TFEU. Article 196(1) TFEU provides that "the

---

281 See Calliess, 2190.

282 Article 196 TFEU:

"1. The Union shall encourage cooperation between Member States in order to improve the effectiveness of systems for preventing and protecting against natural or man-made disasters.

Union action shall aim to:

(a) support and complement Member States' action at national, regional and local level in risk prevention, in preparing their civil-protection personnel and in responding to natural or man- made disasters within the Union;

(b) promote swift, effective operational cooperation within the Union between national civil- protection services;

(c) promote consistency in international civil-protection work.

2. The European Parliament and the Council, acting in accordance with the ordinary legislative procedure shall establish the measures necessary to help achieve the objectives referred to in paragraph 1, excluding any harmonisation of the laws and regulations of the Member States."

Union shall encourage cooperation between Member States in order to improve the effectiveness of systems for preventing against natural or man-made disasters". Under Article 196(2) TFEU, the European Parliament and the Council can establish supplementary measures to assist Member States achieving such objectives[283]. It can be clearly derived from the wording of Article 196 TFEU that the remit of this provision is oriented at building up the necessary capacities in the field of civil protection. "In this context, Article 222 TFEU would operate alongside Article 196 TFEU as a means of adopting civil protection legislation."[284] Article 196 and Article 222 TFEU are, therefore, complementary provisions. Bearing in mind the Joint Proposal, the Solidarity Clause could be even more than complementary to the EU policy area 'Civil Protection'. According to Article 5(1)(d) of the Joint Proposal the Commission, and the High Representative may "where appropriate, submit proposals to Council, in particular as concerns: operational decisions on reinforcement of existing mechanisms, decisions on exceptional measures from Member States not foreseen by existing instruments; policy coordination and information exchange with the aim to create necessary regulatory conditions; operational or support measures for fast reaction of Member States"[285]. This provision opens up the possibility to further integrate the policy area 'Civil Protection' through *ad hoc* legislation.

VIII.c. Relation to Article 214 TFEU[286]

Article 214 TFEU establishes humanitarian aid as a separate policy area (formerly within Article 179 TEC). In the field of humanitarian aid policy,

---

283 See Theodore Konstadinides, 'Evidence on the Government's Review of the Balance of Competences between the United Kingdom and the European Union', 1, available at https://www.gov.uk/government/uploads/system/uploads/attachment_data/file/224408/evidence-theodore-konstadinides.pdf.

284 Konstadinides, 'Evidence on the Government's Review of the Balance of Competences between the United Kingdom and the European Union', 4.

285 18124/12, 11.

286 Article 214 TFEU:
"1. The Union's operations in the field of humanitarian aid shall be conducted within the framework of the principles and objectives of the external action of the

the European Parliament, together with the Council of the EU, acts as co-legislator. The Commission's implementation measures are also submitted to the European Parliament, which has oversight powers[287]. Article 214 could be affected by Article 222 TFEU, which does not have any territorial limitation with regard to the protection of the civilian population and democratic institutions (paragraph (1)(a)). Should the Union envisage humanitarian assistance to third countries, such assistance needs to be based on Article 214 TFEU.

Union. Such operations shall be intended to provide ad hoc assistance and relief and protection for people in third countries who are victims of natural or man-made disasters, in order to meet the humanitarian needs resulting from these different situations. The Union's measures and those of the Member States shall complement and reinforce each other.

2. Humanitarian aid operations shall be conducted in compliance with the principles of international law and with the principles of impartiality, neutrality and non-discrimination.
3. The European Parliament and the Council, acting in accordance with the ordinary legislative procedure, shall establish the measures defining the framework within which the Union's humanitarian aid operations shall be implemented.
4. The Union may conclude with third countries and competent international organisations any agreement helping to achieve the objectives referred to in paragraph 1 and in Article 21 of the Treaty on European Union.
   The first subparagraph shall be without prejudice to Member States' competence to negotiate in international bodies and to conclude agreements.
5. In order to establish a framework for joint contributions from young Europeans to the humanitarian aid operations of the Union, a European Voluntary Humanitarian Aid Corps shall be set up. The European Parliament and the Council, acting by means of regulations in accordance with the ordinary legislative procedure, shall determine the rules and procedures for the operation of the Corps.
6. The Commission may take any useful initiative to promote coordination between actions of the Union and those of the Member States, in order to enhance the efficiency and complementarity of Union and national humanitarian aid measures.
7. The Union shall ensure that its humanitarian aid operations are coordinated and consistent with those of international organisations and bodies, in particular those forming part of the United Nations system."

287 See the fact sheet of the European Parliament, 'Humanitarian Aid', available at http://www.europarl.europa.eu/ftu/pdf/en/FTU_6.5.3.pdf.

Article 214 TFEU would be the *lex specialis* with regard to the assistance of the civilian population in third countries.

VIII.d. Relation to Articles 346–347 TFEU

Article 346 TFEU[288] introduces a derogation from EU law on the basis of public security[289]. The same goes for Article 347 TFEU[290], which sets out a general safeguard clause[291]. The provisions already existed before the entry into force of the Lisbon Treaty but did not apply to the CFSP[292]. These provisions are to be exercised very restrictively in particular as regards Article 347 TFEU, which can only be exercised subsidiary to other exceptional

---

288 Article 346 TFEU:
"1. he provisions of the Treaties shall not preclude the application of the following rules:
(a) no Member State shall be obliged to supply information the disclosure of which it considers contrary to the essential interests of its security;
(b) any Member State may take such measures as it considers necessary for the protection of the essential interests of its security which are connected with the production of or trade in arms, munitions and war material; such measures shall not adversely affect the conditions of competition in the internal market regarding products which are not intended for specifically military purposes.
2. The Council may, acting unanimously on a proposal from the Commission, make changes to the list, which it drew up on 15 April 1958, of the products to which the provisions of paragraph 1(b) apply."

289 Panos Koutrakos, 'The EU Common Security and Defence Policy', (Oxford: Oxford University Press [2013]), 253.

290 Article 347 TFEU:
"Member States shall consult each other with a view to taking together the steps needed to prevent the functioning of the internal market being affected by measures which a Member State may be called upon to take in the event of serious internal disturbances affecting the maintenance of law and order, in the event of war, serious international tension constituting a threat of war, or in order to carry out obligations it has accepted for the purpose of maintaining peace and international security."

291 See Calliess, 2710.

292 See Hafner, ‚Beistands- und Solidaritätsklausel', 396.

provisions[293]. It can be questioned whether Article 347 TFEU could be invoked in order to deny the obligations arising out of Article 42(7) TEU and Article 222 TFEU. The exceptions covered by Articles 346 and 347 TFEU have so far been interpreted as if they could not be invoked if the Community (now Union) had already exercised its competence[294]. However, neither the Mutual Assistance nor the Solidarity Clauses have been applied yet, which opens up the possibility for a Member State to refer to the exceptional provision in Article 347 TFEU[295]. Neutral or non-aligned Member States can also invoke the Irish Clause in Article 42(7) TEU, which raises the question of which of the provisions should prevail. In this respect it is to be noted that the Irish Clause is wider than Article 347 TFEU[296]. As regards the third element of Article 347 ("obligations […] accepted for the purpose of maintaining peace and international security"), this exception is absorbed by the Irish Clause. With regard to Article 222 TFEU there is no exception for neutral and non-aligned Member States. Therefore, in a situation where the particular security policy of these Member States is affected by the Solidarity Clause, the invocation of Article 347 TFEU would be possible. Certainly, non-compliance with the obligations arising out of Article 42(7) TEU and Article 222 TFEU would be subject to scrutiny by the CJEU.

293 Cf the references to the jurisdiction of the ECJ (Leifer and others) in Calliess, 2710.
294 See Hafner, ‚Beistands- und Solidaritätsklausel', 396.
295 Ibid.
296 Ibid., 397.

IX. Concluding Remarks

The Mutual Assistance Clause and the Solidarity Clause are similar and at the same time quite diverse provisions. They are similar because both regulations are binding on all Member States who are obliged to act in a spirit of solidarity and their scope is, in principle, sufficiently broad and flexible to generate possible overlaps. Both Clauses do not limit the assistance nor the means that can be applied to assist the affected Member State(s). Despite these similarities they are separate in legal terms as the one Clause has a clear intergovernmental background whereas the other is supranational. Certainly, the supranationality of the Solidarity Clause does not prevent it from potentially extending its scope to clearly intergovernmental areas such as defence and foreign policy. The intergovernmentalism of the Mutual Assistance Clause is absolute as no involvement of Union institutions is foreseen, whereas with respect to the Solidarity Clause the Union is not only a direct addressee of the solidarity obligation but has also a role (though a marginal one) as regards the Member States' efforts in providing support to the affected Member State(s). Non-compliance with the obligation to assist arising out of Article 42(7) TEU does not result in any legal consequences. What is more, mandatory assistance only binds the Member States which are also NATO members, whereas the neutral and non-aligned Member States are by and large (but not completely, as outlined above) exempted from any obligation to assist. The Member State(s) that decide(s) not to follow the obligation established by the Solidarity Clause could incur infringement proceedings before the CJEU. Furthermore, with respect to procedural issues, both Clauses are distinct: the Mutual Assistance Clause does not prescribe any procedure to be followed in the event of armed aggression; the Solidarity Clause refers to arrangements for the implementation of the Clause by the Union which are still awaiting their final adoption. Even with regard to the assistance by the Member States the Solidarity Clause renounces limiting the freedom of the Member States and provides for a forum (the Council) where they are to coordinate themselves.

The impact of the Mutual Assistance and the Solidarity Clauses can hardly be estimated as both Clauses have not yet been invoked. Whether one or the other Clause, or even both, can contribute to further European integration and a more coherent and interlinked security policy of the Union cannot be predicted. There are, however, some reasons to query whether the Clauses will ever have any impact on European integration at all. Certainly, Europeanisation has always been unsteady and unpredictable, but as long as the main policy areas affected by the Clauses, namely foreign affairs, defence, civil protection, police cooperation, etc., remain as heterogenous and hardly interlinked as they are today, both legally and institutionally, there is not much hope that the Clauses will contribute much to a more integrated security policy. One of the main problems continues to be the pillar structure of the Union, which has been only partially abolished by the Treaty of Lisbon. Despite the countless declarations of intent of the EU institutions and the Member States to surmount the division between internal and external security and to enhance cooperation between the CFSP (and CSDP) and in particular the Area of Freedom, Security and Justice, progress remains modest. And yet, the CFSP and CSDP would probably be the areas where supranational EU action would prove most effective and find the support of the EU citizens. An EU of 28 different EU external security and defence policies, 28 different armies and defence procurement policies is more than ever challenged by the current financial and economic crises. Furthermore, the "Member States, which see their common defence realised in the North Atlantic Treaty Organisation" (Art. 42(2) TEU) could sooner or later be confronted with a situation where the United States of America becomes self-sufficient as regards their energy supply (in particular oil and gas)[297] and, therefore, may reduce its military commitment in different

297 "The [...] US is dealing with an historical energy revolution, the ripples of which can also be felt around the world. Recent developments in the exploitation of non-conventional oil and gas, especially shale, could lead the country to energy independence in the short term. By 2020, the US could be the largest oil and gas producer in the world, surpassing even Saudi Arabia and Russia. By 2030 the country could be a net exporter. In 2005, 60% of oil in the US was imported; that figure already fell to 39% last year. While the stock of regular natural gas in the US decreases, shale gas

parts of the world and, as a consequence, also in NATO. As the United States is by far the biggest 'shareholder' in NATO, European NATO Members could be forced to reconsider their reluctance towards a common European defence policy. Armed aggression against a Member State will, hopefully, never happen but in the unfortunate event that it does (and if we consider cyber attacks of a certain gravity armed aggression according to Art. 42(7) TEU such a case will probably occur), it could trigger the necessary momentum to establish an integrated collective defence system of the Union. However, due to the still predominant role of NATO for the security and defence policies of most Member States, it is rather unlikely that the Mutual Assistance Clause will become the object of particular attention from both scholars and politicians in the near future.

From a legal point of view Article 42(7) TEU is less challenging than the Solidarity Clause. The Irish Clause, however, is worth reconsidering in the light of Article 222 TFEU, as well as of the 'Protocol on the concerns of the Irish people on the Treaty of Lisbon'. The perpetrators of a terrorist attack cannot always be clearly recognized when the attack occurs and it could happen that a neutral Member State provides assistance to a Member State affected by a State-sponsored attack. An *ex post* withdrawal of the assistance after the identification of the perpetrators would be difficult to argue on the basis of the Irish Clause. Though this case might seem merely hypothetical, it shows us that the possible effects of the application of the Solidarity Clause could influence the meaning of the Irish Clause.

The Solidarity Clause is certainly one of the most interesting innovations of the Treaty of Lisbon. This "provision creates one of the most explicit demands upon EU members to act jointly and to assist one another in the face of disasters, emergencies, and crises on the European continent"[298]. The Solidarity Clause can thus be seen "as a treaty-based method for improving EU cooperation

provides 40% of domestic needs. Twelve years ago it was 2%", (Alma de Walsche, 'Suspicious energy? The American shale gas revolution', 4 June 2013, available at http://www.mo.be/en/article/suspicious-energy).

298 Myrdal, Rhinard, 3.

on a range of complex threats"[299], which would have suggested an earlier and more intense analysis with regard to the precise meaning of its solidarity obligation and its implications for EU institutions and Member States by the Commission and the Council. Unfortunately, the Commission and the High Representative tabled a Joint Proposal on the implementing arrangements for the Union that is at the same time vague and timid, but also disappoints most Member States who are looking for a stronger role for the Council. The negotiations of the Joint Proposal will show if "more ambitious arrangements could be so adopted in order to establish an efficient transnational framework and to ensure coherence and visibility of policy decisions, especially vis-à-vis EU citizens"[300]. The Joint Proposal should, on the one hand, highlight the meaning of the Solidarity Clause for European integration and encourage its invocation in exceptional crisis situations but, on the other hand, it should not facilitate the misuse of the Clause for minor incidents or jeopardize existing instruments like the 'Civil Protection Mechanism'. The discussions on the Joint Proposal within the framework of the Council also show one of the main challenges of the Solidarity Clause: 'solidarity' means different things to different people. What is more, this "could lead to conflicting expectations and possible misunderstandings"[301]. Therefore, it is more necessary than ever to establish clear procedures for the solidarity obligations of the Union which, however, cannot be regarded as detached from the obligations of the Member States. "Concerned about making themselves liable before the European Court of Justice for any failure to show solidarity"[302], the Member States may be reluctant to clarify the meaning of Article 222 TFEU. However, in most cases it will be the Member States which have the proper means to respond to emergency situations and, hence, should be entitled through the Council to play a major role in the implementation of the Solidarity Clause by the Union. It is, therefore, in the interest of all Member States to establish clear rules and procedures for the implementation of the Solidarity Clause.

---

299 Ibid.

300 Fuchs-Drapier, 12.

301 Myrdal, Rhinard, 16.

302 Ondarza, Parkes, 8.

Our brief analysis has also shown that the Solidarity Clause could create a certain momentum for a possible widening of the Union's competence, particularly as regards the prevention of terrorist attacks and terrorist threats. It cannot be derived from the *travaux préparatoires* what the authors of Article 222 TFEU had in mind when drafting this provision. There is no doubt that the Solidarity Clause in general, and its preventive scope in particular, challenge the Member States' sovereignty. The competence to take preventive action could also question the capability of certain Member States to fulfil their duties with respect to providing for national security and public order. What if a Member State manifestly ignores the terrorist threat and fails to take necessary measures to prevent terrorist attacks? Should the Union remain silent as long as the affected Member State does not invoke the Clause? Should the other Member States accept in the name of sovereignty that innocent people are killed because of the negligence of its political authorities or should the Member States take preventive action?

"From a political point of view, allowing the triggering of the Clause from a non-affected Member State will reveal that EU solidarity in view of collective security can prevail over national sovereignty."[303] Sovereignty should not only be seen as a right but also as a duty of the Member States to keep high standards as regards their prevention, resilience and response capabilities, regardless of the threat. The implementing arrangements are not to be regarded by some Member States as a sort of *carte blanche* for the reduction of their own capabilities. On the contrary, if well drafted, the Solidarity Clause's implementing arrangements could contribute to higher standards in the field of civil protection and counter-terrorism. We should bear in mind that the European Union is a community not only of shared values, but also of shared threats and risks. The negligence of one Member State can result in a threat to other Member States' populations. This applies to the situations covered by the Mutual Assistance and the Solidarity Clauses but – as can be seen with the current economic and financial crisis in the Eurozone – also to other areas. Therefore, also a sense of

303 Fuchs-Drapier, 4.

responsibility, not only towards their own citizens but also towards the citizens of other Member States, should prevail over traditional and – at least with respect to "an ever closer"[304] European Union – somehow outdated concepts of sovereignty.

304 Preamble of the TEU.

X. Bibliography

a. Literature

Ian Brownlie, 'Principles of Public International Law' (Oxford: Oxford University Press [2008]).

Christian Calliess, ‚Art. 222 AEUV - Solidaritätsklausel', in Christian Calliess, Matthias Ruffert (Hrsg.), ‚EUV/AEUV – Das Verfassungsrecht der Europäischen Union mit Europäischer Grundrechtecharta' (München: Beck [2011]).

Chinchón Alvarez, J. y Sánchez-Bayón, ‚Comentario a los artículos I-43, III-284 y III-329 del Tratado por el que se establece una Constitución para Europa. Los artículos I-43 y III-329: Demasiadas puertas abiertas', in Brú Purón, Carlos Maria (director), ‚Exégesis Conjunta de los Tratados Vigentes y Constitucional Europeos' (Thomson-Civitas [2005]), available at http://eprints.ucm.es/17550/1/Los_art%C3%ADculos_I-43_y_III-329-Demasiadas_puertas_abiertas.pdf.

Paul Craig, Gráienne de Búrca, 'EU Law – Text, Cases and Materials' (Oxford: Oxford University Press [2011]).

Gerhard Hafner, ‚Beistands- und Solidaritätsklausel', in Waldemar Hummer, Walter Obwexer, ‚Der Vertrag von Lissabon' (Baden-Baden: Nomos Verlag [2009]).

Gerhard Hafner, 'The Definition of the Crime of Terrorism', in Giuseppe Nesi, 'International Cooperation in Counter-terrorism' (Ashgate Publishing [2006]).

Waldemar Hummer, ‚Paradigmenwechsel in der „Zweiten Säule" der Europäischen Union. Von der „WEU-geführten" zur „EU-geführten" Sicherheits- und Verteidigungspolitik', in Michael Gehler, Anton Pelinka, Günter Bischof (Hg.), ‚Österreich in der Europäischen Union. Bilanz seiner Mitgliedschaft' (Wien: Böhlau Verlag [2003]).

Panos Koutrakos, 'The EU Common Security and Defence Policy' (Oxford: Oxford University Press [2013]).

Araceli Mangas Martín, Diego J. Liñán Nogueras, ‚Institutiones y Derecho de la Unión Europea' (Madrid: Editorial Tecnos [2012]).
Peter Part, ‚Artikel 120–126 AEUV', in Heinz Mayer, Karl Stöger, ‚Kommentar zu EUV und AEUV unter Berücksichtigung der österreichischen Judikatur und Literatur' (Wien: Manz Verlag [2012]), (30).
Steve Peers, 'EU Justice and Home Affairs Law' (Oxford: Oxford University Press [2011]).
August Reinisch, 'The Action of the European Union to Combat International Terrorism', in Andrea Bianchi, 'Enforcing International Law Norms Against Terrorism'(Oxford and Portland Oregon [2004]).
Michael Rodi, ‚Artikel III-180', in Christoph Vedder, Wolff Heintschel von Heinegg, ‚Europäischer Verfassungsvertrag' (Baden-Baden: Nomos Verlag [2007]).
Gregor Schusterschitz, Georg Stillfried, 'Art 222 AEUV', in Heinz Mayer, Karl Stöger, ‚Kommentar zu EUV und AEUV unter Berücksichtigung der österreichischen Judikatur und Literatur' (Wien: Manz Verlag [2012]).
Christoph Vedder, ‚Artikel I-41', in Christoph Vedder, Wolff Heintschel von Heinegg, ‚Europäischer Verfassungsvertrag' (Baden-Baden: Nomos Verlag [2007]).

b. Articles

Bruno Angelet, Ioannis Vrailas, 'European Defence in the wake of the Lisbon Treaty', (Egmont Royal Institute, Egmont Paper 21 [May 2008]).
Michael J. Austin, 'Style or Substance? An Analysis of the Major Reforms to CFSP by the Treaty of Lisbon' (European Union Law Working Papers No. 1, Stanford/Vienna Transatlantic Technology Law Forum [2011]).
Karen Devine, 'Irish Neutrality and the Lisbon Treaty', available at http://doras.dcu.ie/14898/1/Irish_Neutrality_and_the_Lisbon_Treaty.pdf.
Dolores Rubio García, ‚Las cláusulas de asistencia mutual y solidaridad introducidas por el Tratado de Lisboa: el refuerzo de la seguridad y la defensa en la Unión Europea' (Documento de Trabajo 57/2011).
Sofia Fernandes, Eulalia Rubio, 'Solidarity within the Eurozone: how much, what for, for how long?' (Notre Europe Policy Paper No. 51 [2012]),

available at http://www.notre-europe.eu/media/SolidarityEMU_S.Fernandes-E.Rubio_NE_Feb2012.pdf.

Laura C. Ferreira-Pereira and A. J. R. Groom, "Mutual solidarity' within the EU common foreign and security policy: What is the name of the game?' (International Politics [2010] 47).

Yoram Dinstein, 'Computer Network Attacks and Self-Defense' (76 INT'L L. STUD. 99 [2002]). Cited by Matthew C. Waxman, 'Cyber-Attacks and the Use of Force: Back to the Future of Article 2(4)' (THE YALE JOURNAL OF INTERNATIONAL LAW [Vol. 36]), available at http://www.yjil.org/docs/pub/36-2-waxman-cyber-attacks-and-the-use-of-force.pdf.

Marie Fuchs-Drapier, 'The European Union's Solidarity Clause in the Event of a Terrorist Attack: Towards Solidarity or Maintaining Sovereignty?' (Blackwell Journal of Contingencies and Crisis Management, Volume 19 Number 4 [December 2011]).

Eduardo Gianfrancesco, ‚La Politica Estera e di Sicurezza Europea prima e dopo il Trattato di Lisbona' (Forum di Quaderni Costituzionali [2011]), available at http://www.forumcostituzionale.it/site/images/stories/pdf/documenti_forum/paper/0267_gianfrancesco.pdf.

Myrto Hatzigeorgopoulos, 'The Mutual Assistance and Solidarity Clauses' (ISIS Europe European Security Review 61 [2012]).

Peter Hilpold, ‚Österreichs Neutralität nach Lissabon im Lichte der Beistands- und Solidaritätsverpflichtung' (ÖJZ 2010/65).

Jolyon Howorth, 'The European Draft Constitutional Treaty and the Future of the European Defence Initiative: A Question of Flexibility' (European Foreign Affairs Review 9 [2004]), available at http://www.cap.lmu.de/transatlantic/download/howorth.pdf.

Chris Jones, 'Secrecy reigns at the EU's Intelligence Analysis Centre' (Statewatch Analysis), available at http://www.statewatch.org/analyses/no-223-eu-intcen.pdf.

Jacques Keller-Noellet, 'The Solidarity Clause of the Lisbon Treaty's', available at http://www.notre-europe.eu/media/tgae20117fkellernoellet.pdf?pdf=ok.

Theodore Konstadinides, 'Civil Protection in Europe and the Lisbon 'solidarity clause': A genuine legal concept or a paper exercise' (Uppsala Faculty of Law, Working Paper [2011]), available at http://www2.statsvet.uu.se/LinkClick.aspx?fileticket=ywEQQ722UuI%3D&tabid=3159&language=sv-SE.

Sara Myrdal, Mark Rhinard, 'The European Union's Solidarity Clause: Empty Letter or Effective Tool? An Analysis of Article 222 of the Treaty on the Functioning of the European Union' (Swedish Institute of International Affairs [2010]).

Nicolai von Ondarza, Roderick Parkes, ‚The EU in the face of disaster' (Stiftung Wissenschaft und Politik, SWP Comments 9 [April 2010]).

Steve Peers, 'Analysis - Lisbon Treaty guarantees for Ireland, 19 June 2009', available at http://www.statewatch.org/news/2009/jun/lisbon-ireland.pdf, 1.

Jochen Rehrl, ‚Beistandsgarantie und Solidaritätsklausel. Völker- und Verfassungsrechtliche Herausforderungen für Österreich durch den neuen EU-Verfassungsvertrag' (Zeitschrift für öffentliches Recht 60 [2005]).

Abraham D. Sofaer, David Clark, Whitfield Diffie, 'Cyber Security and International Agreements', in 'Proceedings of a workshop on deterring Cyberattacks: Informing strategies and developing options for U.S. policy (2010)', available at http://www.nap.edu/openbook.php?record_id=12997&page=185.

Sophie Vanhoonacker, 'Common Foreign and Security Policy: Can History be overcome?' (Paper Prepared for ECSA Conference, Seattle [May 1997]), available at http://aei.pitt.edu/6838/1/002722.PDF.

Alma de Walsche, 'Suspicious energy? The American shale gas revolution', 4 June 2013, available at http://www.mo.be/en/article/suspicious-energy).

c. Documents of the Council of the European Union

Austrian Delegation, 'Implementation of Article 222 TFEU (Solidarity Clause)' (10956/12), available at http://www.statewatch.org/news/2013/feb/eu-solidarity-clause-austria-10956–12.pdf http://www.statewatch.org/news/2013/feb/eu-solidarity-clause-austria-10956–12.pdf.

Contribution of the Council Legal Service: Joint proposal for a Council Decision on the arrangements for the implementation by the Union of the Solidarity Clause (Article 222 TFEU) - Scope (14498/13).

Coordinating Committee in the area of police and judicial cooperation in criminal matters (ex- 'Article 36 Committee' or 'CATS') (16070/09).

Council of the European Union IPCR Secretariat, 'IPCR in brief' (*non-official Council document*).

'Council Regulation (EC) 2012/2002 of 11 November 2002 establishing the European Union Solidarity Fund' [2002] OJ L 311/3.

'Council Decision 2007/779/EC, Euratom of 8 November 2007 establishing a Community Civil Protection Mechanism (recast)' [2007] OJ L 314/9.

'Council Regulation (EU) No 407/2010 of 11 May 2010 establishing a European financial stabilization mechanism' [2010] OJ L 118/1.

'Finalisation of the CCA review process: the EU Integrated Political Crisis Response (IPCR) arrangements' (10708/13).

'First main assessment and description report for internal debate (M.A.D.R.I.D. Report)' (10203/10).

'Framework Decision of the Council of 13 June 2002 on combating terrorism (2002/475/JHA)' [2002] OJ L 164/4.

'Handling within the Council of the joint proposal by the Commission and the High Representative on the arrangements for the implementation by the Union of the Solidarity Clause (Art. 222 TFUE)' (6598/13).

'Joint proposal for a Council Decision on the arrangements for the implementation by the Union of the Solidarity Clause' (18124/12).

'Press Release 2732nd Council Meeting' (9409/06, Presse 144).

'Press Release 3279th Council Meeting' (17342/13, Presse 534).

'Solidarity Clause – results of the orientation debate on Art. 222 TFEU at the CATS meeting on 6 October 2011' (17185/11).

'The European Union Counter-Terrorism Strategy' (14469/4/05 Rev 4).

'The European Union Strategy for Combating Radicalisation and Recruitment to Terrorism' (14781/1/05).

'The Internal security strategy for the European Union - Towards a European security model' (5842/2/10).

d. Documents of the European Commission

Communication from the Commission to the European Parliament and the Council of 22 November 2010, 'The EU Internal Security Strategy in Action: Five steps towards a more secure Europe' (COM(2010) 673 final).

European Commission, 'Civil protection legal texts', available at http://ec.europa.eu/echo/civil_protection/civil/prote/legal_texts.htm

European Commission, 'Emergency Response Coordination Centre (ERCC)', available at http://ec.europa.eu/echo/about/ERC_en.htm.

'Proposal for a decision of the European Parliament and of the Council on serious cross-border threats to health' (COM(2011) 866 final).

e. Documents of the European Parliament

European Parliament resolution of 22 November 2012 on the EU's mutual defence and solidarity clauses: political and operational dimensions (2012/2223(INI)).

Fact sheet of the European Parliament, 'Humanitarian Aid', available at http://www.europarl.europa.eu/ftu/pdf/en/FTU_6.5.3.pdf.

f. Documents of the European Council

'A secure Europe in a better World - European Security Strategy' (12 December 2003).

'Conclusions of the European Council (19/20 December 2013)' (EUCO 217/13).

'Declaration on solidarity against terrorism' (25 March 2004).

'Presidency Conclusions - European Council Meeting 14 and 15 December 2001 in Laeken' (SN 300/1/01).

g. Treaties

'North Atlantic Treaty' of 4 April 1949, 34 UNTS 243–246, available at http://www.nato.int/docu/basictxt/treaty.htm.

Official Journal of the European Union C 115/13, 'Consolidated Version of the Treaty on the European Union', 9.5.2008.
Official Journal of the European Union C 326/1, 'Consolidated Versions of the Treaty on European Union and the Treaty on the Functioning of the European Union', 26.10.2012.
Official Journal of the European Union C 169/3, 'Draft Treaty establishing a Constitution for Europe', 18.7.2003.
Official Journal of the European Union L 60/131, 'Protocol on the concerns of the Irish people on the Treaty of Lisbon', 2.3.2013.
'Protocol Additional to the Geneva Conventions of 12 August 1949, and relating to the protection of victims of international armed conflicts (Protocol I)', UNTS I-17512.
'Single European Act', UNTS I-30614.
'Treaty between Belgium, France, Luxembourg, Netherlands and the United Kingdom of Great Britain and Northern Ireland for collaboration in economic, social and cultural matters and for collective self-defence', UNTS I-304.
'Treaty of Lisbon amending the Treaty on European Union and the Treaty establishing the European Community', UNTS I-47938.
'Vienna Convention on the Law of Treaties' of 23 May 1969, UNTS I-18232.

h. Documents of the European Convention and the IGC 2003/2004

'Comments by Mr Manuel Lobo Antunes on the preliminary draft final report of Working Group VIII "Defence" (WD 022)'.
'Comments by Mrs Maria Eduarda Azevedo on the preliminary draft final report of Working Group VIII "Defence" (WD 022)'.
'Comments by Mr Slavko Gaber on the preliminary draft final report of Working Group VIII "Defence" (WD 022)'.
'Comments by Mr John Gormley on the preliminary draft final report of Working Group VIII "Defence" (WD 022)'.
'Comments by Ms Danuta Hubner on the preliminary draft final report of Working Group VIII "Defence" (WD 022)'.

‘Comments by Ms Sylvia-Yvonne Kaufmann on the preliminary draft final report of Working Group VIII “Defence” (WD 022)’.
‘Comments by Mr Kimmo Kiljunen to the preliminary draft final report of Working Group VIII “Defence” (WD 022)’.
‘Comments by Mr Kenneth Kvist on the preliminary draft final report of Working Group VIII “Defence” (WD 022)’.
‘Comments by Mr Esko Seppänen on the preliminary draft final report of Working Group VIII “Defence” (WD 022)’.
‘Comments by Ms Gisela Stuart to the preliminary draft final report of Working Group VIII “Defence” (WD 022)’.
‘Draft Treaty Establishing a Constitution for Europe’ (CONV 850/03).
‘Dutch comments on the preliminary draft final report of Working Group VIII “Defence” (WD 022)’.
‘Final report of Working Group VIII – Defence’ (CONV 461/02).
‘Franco-German comments on the preliminary draft final report of Working Group VIII “Defence” (WD 022)’.
‘IGC 2003 - European Security and Defence Policy’ (CIG 62/03)
‘Mandate of the Working Group on Defence’ (CONV 206/02 and CONV 246/02).
‘Preliminary draft report from the Working Group on Defence (WD 022)’.

i. Other Sources

Michel Barnier, ‘For a European civil protection force: europe aid’, available at http://ec.europa.eu/archives/commission_2004-2009/president/pdf/rapport_barnier_en.pdf.
Andy Brown, ‘Ten Years of Europol 1999–2009’, available at https://www.europol.europa.eu/sites/default/files/publications/anniversary-publication.pdf, 25.
Draft Articles on the Law of Treaties with commentaries, available at http://untreaty.un.org/ilc/texts/instruments/english/commentaries/1_1_1966.pdf, 37.
Joint Statement of the Heads of State and Government of Germany, France, Luxembourg and Belgium on European Defence, Brussels, 29 April 2003

(“Tervuren Declaration”), available at http://www.ambafrance-uk.org/Meeting-of-the-Heads-of-State-and.html.

ICJ, Reports of Judgments, Advisory opinions and orders, ‘Case concerning military and paramilitary activities in and against Nicaragua (Nicaragua v. United States of America)’.

Institut de Droit International (Institute of International Law), ‘Resolution on “Humanitarian Assistance”’ (Bruges 2003 session), available at http://www.idi-iil.org/idiE/resolutionsE/2003_bru_03_en.PDF.

Theodore Konstadinides, ‘Evidence on the Government’s Review of the Balance of Competences between the United Kingdom and the European Union’, available at https://www.gov.uk/government/uploads/system/uploads/attachment_data/file/224408/evidence-theodore-konstadinides.pdf.

Statement of the first director of Ilkka Salmi, Director of the Situation Centre of the European Union, available at http://www.europarl.europa.eu/meetdocs/2009_2014/documents/sede/dv/sede041011cvsalmi_/sede041011cvsalmi_en.pdf.

‘The United Kingdom’s Strategy for Countering Terrorism’ published in July 2011 which keeps its typical structure (Pursue, Prevent, Protect, Prepare), available at https://www.gov.uk/government/uploads/system/uploads/attachment_data/file/97994/contest-summary.pdf.

UN General Assembly Resolution A/RES/3314(XXIX) of 14 December 1974.

UN Security Council Resolution S/RES/1368 (2001) of 12 September 2001.

j. Table of Cases

Commission vs. Italy (Case 39/72) [1973] ECR 101.

Maizena vs. Council (Case 139/79) [1980] ECR 3393.

Roquette Frères vs. Council (Case 138/79) [1980] ECR 3333.